GREWINGK'S GEOLOGY OF ALASKA
AND THE NORTHWEST COAST OF AMERICA

RASMUSON LIBRARY TRANSLATION SERIES

Marvin W. Falk, Editor

1. *Holmberg's Ethnographic Sketches* by Heinrich Johan Holmberg. Edited by Marvin W. Falk, translated from the original German of 1855–1863 by Fritz Jaensch. 1985.

2. *Tlingit Indians of Alaska* by Anatolii Kamenskii. Translated, with an introduction and supplementary material, by Sergei Kan. 1985.

3. *Bering's Voyages: The Reports from Russia* by Gerhard Friedrich Müller. Translated, with commentary, by Carol Urness. 1986.

4. *Russian Exploration in Southwest Alaska: The Travel Journals of Petr Korsakovskiy (1818) and Ivan Ya. Vasilev (1829)*. Edited, with an introduction by James VanStone, translated by David H. Kraus. 1988.

5. *The Khlebnikov Archive: Unpublished Journal (1800–1837) and Travel Notes (1820, 1822, and 1824)*. Edited by Leonid Shur, translated by John Bisk. 1990.

6. *The Great Russian Navigator, A. I. Chirikov* by Vasilii A. Divin. Translated and annotated by Raymond H. Fisher. 1993.

7. *Journals of the Priest Ioann Veniaminov in Alaska, 1823 to 1836*. Introduction and commentary by S. A. Mousalimas, translated by Jerome Kisslinger. 1993.

8. *To the Chukchi Peninsula and the Tlingit Indians 1881–1882: Journals and Letters* by Aurel and Arthur Krause. Translated by Margot Krause McCaffrey. 1993.

9. *Essays on the Ethnography of the Aleuts* by R. G. Liapunova. Translated by Jerry Shelest with W. B. Workman and Lydia Black. 1996.

10. *Fedor Petrovich Litke* by A. I. Alekseev. Edited by Katherine L. Arndt, translated by Serge LeComte. 1996.

11. *Grewingk's Geology of Alaska and the Northwest Coast of America: Contributions Toward Knowledge of the Orographic and Geognostic Condition of the Northwest Coast of America, with the Adjacent Islands* by Constantine Grewingk. Edited by Marvin W. Falk, translated by Fritz Jaensch. 2003.

12. *Steller's History of Kamchatka: Collected Information Concerning the History of Kamchatka, Its Peoples, Their Manners, Names, Lifestyle, and Various Customary Practices* by Georg Wilhelm Steller. Translated by Margritt Engel and Karen Willmore. 2003.

13. *Through Orthodox Eyes: Russian Missionary Narratives of Travels to the Dena'ina and Ahtna, 1850s–1930s*. Translated with an introduction by Andrei A. Znamenski. 2003.

Grewingk's Geology of Alaska and the Northwest Coast of America

Contributions Toward Knowledge of the Orographic and Geognostic Condition of the Northwest Coast of America, with the Adjacent Islands

DR. CONSTANTINE GREWINGK

Edited by Marvin W. Falk
Translated by Fritz Jaensch

UNIVERSITY OF ALASKA PRESS
Fairbanks, Alaska

Printed in the United States of America
This publication was printed on acid-free paper that meets the minimum requirements for
American National Standards for Information Sciences—Permanence of Paper for Printed
Library Materials, ANSI z39.48-1984.

LIBRARY OF CONGRESS CATALOGUING-IN-PUBLICATION DATA

Grewingk, C. (Constantin), 1819–1887.
   [Beitrag zur Kenntniss der orographischen und geognostischen Beschaffenheit der Nord-West-
   Küste Amerikas mit den anliegenden Inseln. English]
   Grewingk's geology of Alaska and the northwest coast of America: contributions toward
   knowledge of the orographic and geognostic condition of the northwest coast of America,
   with the adjacent islands / by C. Grewingk; edited by Marvin W. Falk; translated by Fritz
   Jaensch.
      p. cm. — (The Rasmuson Library historical translation series ; v. 11)
   Includes bibliographical references and index.
   ISBN 1-889963-48-8 (pbk. : alk. paper)
   1. Geology—Alaska. 2. Geology—Northwest Coast of North America. 3. Physical geography—Alaska. 4. Physical geography—Northwest Coast of North America.
   I. Title: Geology of Alaska and the northwest coast of America. II. Title. III. Series.
   QE83 .G7413 2002
   557.98—dc21   2002007836

Publications of the Russian Imperial Mineralogical Society at St. Petersburg.
Volumes for 1848 and 1849
St. Petersburg: Carl Kray, 1850
International Standard Book Number 1-889963-48-8

Publication coordination by Jennifer Robin Collier
Text design by Rachel Fudge
Cover design by Mike Kirk

Contents

Acknowledgments

THIS TRANSLATION PROJECT began in the early 1980s with several years of funding provided by the state through the good offices of Charlie Parr, Brian Rogers, and their colleagues in the Alaska State Legislature. Fritz Jaensch had completed the translation and the manuscript was being prepared for printing by AEIDC (Arctic Environmental Information and Data Center) in Anchorage when budget cuts put the project on hold. Philip Marshall, a geologist, had reviewed the manuscript for its geological terminology. Ginger French, Denise Coté, and especially Katherine Arndt have made a number of helpful editorial suggestions.

Elmer Rasmuson's generous annual donations to the library and the support of the library administration—Susan Grigg and Paul McCarthy—have allowed Grewingk's text to be taken up again. The University of Alaska Press has taken over typesetting, book design, printing, and distribution.

Editor's Introduction

CONSTANTINE CASPAR ANDREAS GREWINGK was a leading geologist and minerologist in Russia during the middle of the nineteenth century. During his lifetime he published about 180 articles and several books, with an emphasis on the geology, mineralogy and archeology of the Pribaltic region. He also made major contributions on the geology of the White Sea coast and the Urals. On the way, he took the time to publish a most useful compilation of information on the geology, mineralogy and physical geography of the coast north of San Francisco.

Grewingk graduated from Dorpat University (Tartu, Estonia) and pursued a graduate education at Jena, in Germany, where he received his doctorate. He then became the curator of mineralogical collections at the Academy of Sciences in St. Petersburg in 1845. It was there that he was able to examine the collections of mineral samples sent back by Russian expeditions and various individuals from California, Alaska and the North Pacific. Several years following the publication of this work, he became the librarian of the Mining Institute and then was appointed a professor at his old University of Dorpat in 1854.

This work is the first documented compilation of all of the geologic information about Alaska and adjoining territory available during the Russian-American era. Grewingk himself was never in Alaska, but he exhaustively summarizes the reports and descriptions made by expedition personnel and other direct observers. He presents this information in several useful ways. The main body of the text provides these observations by location, region by region. It serves as something of an index to the published observations by geographic location. He presents a chronology of volcanic events that occurred from Mount St. Helens northward as a table to enhance his textual descriptions. He had access to the mineral collections sent back to the capital from the North Pacific and makes use of them in his presentation. His first appendix is a list of known animal and plant fossils found on the Northwest Coast and the Aleutian Islands. His second appendix is an extended annotated bibliography of the most important published sources for the study of voyages and discoveries arranged chronologically. He includes some sources for discoveries in the American West, including the Lewis and Clark expedition and Fremont.

Grewingk provides a systematic explanation of his understanding of the relationship of the geology of Alaska to that of the rest of North America as well as the Pacific Rim. He illustrates his concept with a drawing and a schematic map. Detailed knowledge of the geology of the American West was in itself in its infancy, so that his explanations are more interesting as an early attempt to make broad sense of geologic history than as anything of current theoretical interest. Still, some of his arguments concerning the development of mountain systems and volcanic activity seem surprisingly modern. He expresses his frustration with the limits of the data at his disposal, and wishes in print that he could be free from the constraints that preclude the printing of the more advanced geological maps that he wanted in his book.

Grewingk's geography supplements and completes Ferdinand Wrangell's *Statistische und ethnographische Nachrichten über die Russischen Besitzungen an der Nordwestküste von Amerika*, 1839 (translated by Mary Sadouski and published by Richard Pierce's Limestone Press as *Russian America: Statistical and Ethnographic Information*, 1980). Wrangell includes extensive observations of the flora and fauna of Alaska but his most significant contribution is the ethnographic description of human populations and the history of Russian occupation of the territory. In contrast, Grewingk's work, nine years later (written in 1848), focuses on geology, not human history.

The value of Grewingk's contribution continued to be relevant for generations following its publication. After the sale of Alaska there were general descriptions of Alaskan geography by William Healy Dall,[1] Ivan Petrof[2] and others. Naturalists made isolated observations, but the systematic study of mineral resources did not begin until Congress began funding the Geological Survey work in Alaska in 1895. Alfred Brooks, head of the United States Geological Survey in Alaska, wrote that Grewingk was "... first to obtain any true conception of the relation of the orographic features of Alaska to those of the rest of North America."[3] Further he noted that this was the "... first attempt at a systematic treatment of the geology, and the results even at the present day are not without value."[4] Grewingk remains even today the best source for observations made from the 1740s to the 1840s.

—*Marvin Falk, Editor*

Notes

1. William H. Dall, *Alaska and its Resources. Boston:* Lee and Shepard, 1870.
2. Ivan Petrov, *Report on the Population, Industries and Resources of Alaska*. Washington, D.C.: GPO, 1884.
3. Alfred H. Brooks, *The Geography and Geology of Alaska: A Summary of Existing Knowledge*. Washington, D.C.: GPO, 1906, p. 201.
4. Ibid.

Preface

The conferences of the Mineralogical Society in St. Petersburg for the year 1847, pp. 142–163, contain several geognostic remarks on Old and New California. They came about as the result of a shipment sent to Mr. Il'ia Voznesenskii, in charge of preparations at the Zoological Museum of the Academy of Sciences at St. Petersburg. Similar, but more comprehensive shipments occasioned subsequent papers, primarily concerning the Russian American colonies.

Admiral Lütke states in one of his excellent travelogues (*Voyage autour du monde, partie nautique*, 1836, p. 279): "L'archipel Aleutien, frequente deja depuis plus d' un siecle par nos batiments et par ceux d'autres nations, est encore aujour d'hui presuq'aussi peu connu qu'au temps de Cook." And this will not surprise anyone who takes to hand one of the more extensive reports of sea voyages in that ocean, nor he who follows, for instance, Glazunov's travelogue in von Baer's and von Helmersen's *Beiträge zur Kenntniss des Russischen Reiches* (Vol. I, pp. 137–161). In it the reader is sufficiently informed about the dangers and hardships of travel in those regions.[1]

Such research is difficult. A thorough investigation in general, as well as a more thorough geologic and geographic one of the entire, immense region, than that which follows, will probably not happen very soon, nor very easily. Therefore, the processing and publishing of each more or less important specimen from there will be a necessary, if thankless, task. Their identification might be not a little welcome to future scientists. But it will have to concur with their independent investigations as based on their own conceptions.[2]

The processing of mineralogic and geologic specimens is made possible by the similarity of organic nature in all climates. We can even draw some common conclusions without our own experience or a close description of a region, simply by studying samples of rocks, minerals, and petrified fragments [fossils] sent to us. The basic rocks of mountain ranges are the same everywhere. There are basalt and lava of an analogous composition and structure everywhere. Even the metamorphic rock formations, investigated less often until recently, we find unaltered, with few exceptions, in the most widely differing regions, regardless of the manifold conditions which brought about their formation.

Humbolt states (in *Kosmos*, I, p. 237):

Where the old stars no longer shine above the seafarer, in islands of far-away oceans, surrounded with plants of strange kinds, he still discerns in details of the landscape's character, as in a mirror image, Mount Vesuvius, the dome-shaped summits of the Auvergne, mountain craters of the Canary Islands and the Azores, and the erupting fissures of Iceland.

An experienced observer can even discern the mineralogic and structural character of a region from outer contours of the land. By the characteristics of the rocks, he can, with much probability, deduce the composition, the distribution of water, and fertility, etc. The use of this method, however, may only be permitted in exceptional cases. It seldom will measure up to scientific requirements. The type of mountain regions where fossils are found permit this method least of all, because the mineralogical characteristics are of little value here. But these fossils do provide us with a sure means for promoting, in a scientific manner, the investigations of our earth's crust, although fossils are only imperfect remnants of organisms, and complete identification of them with known species is often neither provable nor even possible.

CHAPTER ONE

->- -<-

The Western Half of North America Between the Parallel of the Bay of
San Francisco and the Mouth of the Stikine River,
with the Islands Along the Coast

FROM THE NORTHERN INTERIOR of the Bay of San Francisco we see the continuation of the California Coastal Range in several insignificant elevations, extending
northwestward to the coast and, in part, to the northeast along the right bank of
the Sacramento River. The latter range then unites with Mt. Shasta,[3] with a mountain range that lies on the left side of the sources of the river. The Sierra Nevada,
from the western slope of which the gold-containing tributary rivers of the Sacramento come, takes a northeastern direction. And it seems to maintain the same,
without significant interruptions, deeper into the interior of the country.

A short distance away, where the Oregon territory begins, below 41° northern latitude, the division of these mountain regions and of the Rocky Mountains
is not well known. The latter, known also as the Anhuac Range, lose themselves
along this latitude in complicated systems of differently directed mountain
regions, which form major continental divides of water systems between the
Atlantic and Pacific Oceans. Among these systems are the Colorado Range,
Green Range, New Park Range, Sweet Water Range, Medicine Bow Mountains,
Black Hills, and, perhaps also, the Wind River Mountains, from which flow the
rivers that later on gather into the Rio Colorado, which empties into the Gulf of
California. The Snake River, which flows by way of the Columbia River into
the Pacific Ocean, also originates there, and so does the Platte River (tributary
of the Missouri), which finally empties into the Gulf of Mexico.

Along this same latitude, on the western side of the great, high valley, or the
salty sand-plateau,[4] the Sierra Nevada seems to diminish into a hilly region of
insignificant elevation. The Rio Sacramento has its source in the continuation of
the California Coast Range from Mt. Shasta to Mt. Six Cailloux, and it borders
the east shore of Lake Tlamac [Klamath]. From here comes the Klamath River
and Six Cailloux River (tributary of the Columbia River).[5] All three of them
empty into the Pacific Ocean. But their scale is smaller among the less extensive
mountain systems close to the coast. The lack of rivers flowing eastward into the
salty sand-plateau is easily explainable as well.

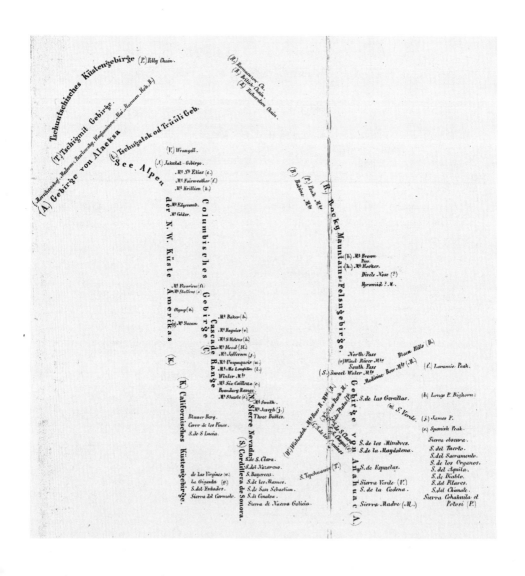

Map 1. Distribution of mountains in the west half of
North America.

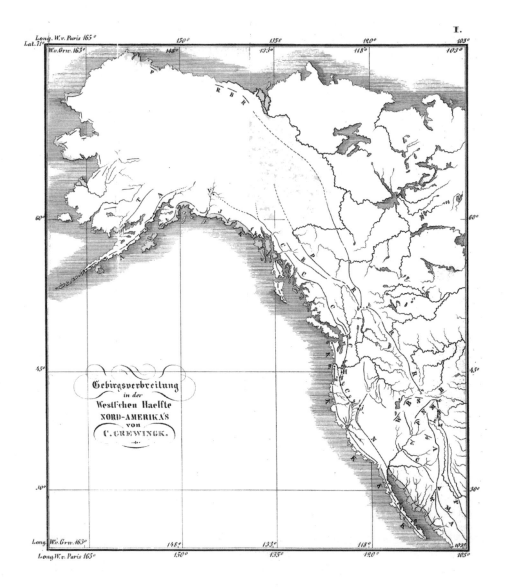

Gebirgsverbreitung
in der
Westl'chen Haelfte
NORD-AMERIKA'S
von
C. GREWINGK.

With the Winter Mountains and the Cascade Range (extending from Mt. Shasta and Mt. Laughlin to Mt. Baker), the Columbia Range has its beginning from Mt. Shasta northward, while the Rocky Mountains join with the Wind River Mountains, after the above-mentioned interruption.

The Columbia or Oregon River breaks through the previously named mountain regions west of the Rocky Mountains together with its tributary, the Snake (Sagaptin or Lewis [Lewis Fork?]) River, which is hardly less sizable than the Columbia River itself. The Columbia has its source in the Wind River Mountains between 43° and 44° N. Lat. The Rocky Mountains sweep almost parallel to the coast of the Pacific Ocean up to about 53° N. Lat. According to Wilkes or Freimann,[6] they feature passes in only three locations, through narrow passageways, between western and eastern North America. The southernmost of them (Fremont's South Pass) actually cuts through the Wind River Mountains between 42° and 43° N. Lat. The middle pass is located between 50° and 53° N. Lat.

From 52° N. Lat., the Rocky Mountains turn westward. And with their highest points, Mt. Hooker (elevation 16,700 ft.) and Mt. Brown (elevation 16,000 ft.), attention is directed to geological events that had taken place on the same parallel along the seacoast. But thereafter the mountain range diminishes quickly in height, as it takes a northwest direction and widens out westward into the region of the Fraser Springs. In this manner it merges with the Columbia Range and comes close to the coast. Along this latitude, the coast forms an almost perpendicular, unscalable rock wall, traversed by steep ravines. A large number of partly barren and partly wooded rock islands are situated along this coast, verifying it as the site of volcanic events.

The regions of the Columbia River (Columbia) and of the Fraser River (New Caledonia), are therefore distinctly separated one from the other by their physical composition. That part of the Columbia between the Rocky Mountains and the Columbia Range which has the character of a desert, has already been described with reference to Upper California. But the Coastal Region, 100–200 nautical miles in width, west of the latter mountains, presents for the most part a friendly, fruitful aspect. Only close to the coast are there mighty, dark forests of conifers, or plains resembling the steppe. The Columbia River itself, after it leaves the constricting rock walls of the Columbia Range,[7] flows along quietly, surrounded everywhere by hills and valleys, with shady woods and fruitful flatlands. At its mouth it is seven miles wide amid an extraordinarily rich, lush region, where hard coal has also been found.

New Caledonia or the Fraser River region, in contrast, has, as a result of its location, its rocky condition, and its volcanic nature, no connection at all with its bordering regions. Northward from the mouth of this river, forced between rock walls, and abounding in waterfalls, only footpaths lead into the interior of the country. The Columbia Range approaches the mouth of the Fraser River after leaving several mighty mountains south of the Columbia River, such as Mt. McLaughlin (43°30' N. Lat., 121° 30' W. Long.), Mt. Jefferson (44°30' N. Lat., 122°30' W. Long.),[8] and Mt. Hood (45°15' N. Lat., 121°30' W. Long.), with an elevation of 7,700 ft. Engl. [English].[9] There are three still-active volcanos: Mt. St. Helens (46°10' N. Lat., 122°15' W. Long.),[10] Mt. Rainier (Regnier, G. Reyner, 47°20' N. Lat.?, 121°50' W. Long.), and Mt. Baker (G. Bekker, 48°45' N. Lat., 121°50' W. Long.). According to Simpson, the first has an elevation of 12,700 ft., and the second 12,500 ft. From then on, this range forms the coast of the mainland (Mt. Steffens, 51° N. Lat., Mt. Fleurieu, 51°30' N. Lat.).

In this manner the arid desert conditions of the region up to the Rocky Mountains are very conspicuous, while the fertile, western part has mostly been filled in by the sea. The entire coast here is crowded with many small islands, which are separated from the mainland by many narrow, dangerous straits. The rocky mountaintops of these islands are covered with snow most of the year, and in the ravines the glaciers almost reach the sea. Frequent fog and storms render the climate harsher yet. And here as well begins the melancholy character of the largest part of the northward-rising coast of the mainland. The inhabitants of the rocky Quadra Islands seem indeed to be related to the Aleuts, judging by their almost white skin color (Erman, l.c., p. 236). Only Charlotte Island is supposed to have completely even terrain and fairly fertile soil.[11]

From what has been said so far, we discern that the region between the lower runs of the Columbia and Fraser Rivers, with its different mountain ranges crowding in on each other, marks the site of volcanic activities, hardly less grandiose than that of Mexico. Farther northward, these ranges do not seem to get any taller. The close proximity of the Rocky Mountains (Peak and Babine Mountains) to the Columbia Range gives a terribly broken appearance to the entire coastal region, that is, the Pacific Coast, with its wealth of islands: New Georgia, New Cornwall, New Hannover, and the Russian American Possessions. Glaciers, which reach down to the sea, fog, and storms make that appearance even more pronounced. This feature also explains the lack of major rivers between 46°30' N. Lat. (Columbia) and 60°15' N. Lat. (Copper River). Known along this distance are only these: the Chickless River, four small rivers that empty into Puget Sound and Admiral's Inlet, and the Fraser River. One river each falls into Bentenick Bay and into Dean Canal. Finally, there are the Salmon River (Simpson, Tacutchee, or Tesse River), the Stachin [Stikine] River, the Taku River, and the Chilkat River. Of these, only the Stikine and Fraser Rivers are of substantial size, but neither of them is navigable.

It is probable that the region between the California Coast Range or Columbia Range and those hills and mountain ridges immediately adjacent to the coast, has in part been shaped by the power of volcanic activities. Upper California is partially covered with volcanic debris, partially by thick coniferous forests, and partially by what appears as steppe. The area around San Francisco Bay has only been lightly affected. From Cape Flattery onward (a part of the Olympia Mountains, 47°45′ N. Lat., 123°24′ W. Long., profiled in Vancouver's *Atlas*, Pl. XIII), complete islands have been formed, such as Quadra Island,[12] Charlotte, Wales,[13] Sitka, etc. Several small islands arose independently from the bottom of the sea. For the large island masses, we need only assume that the rising of the mainland in earlier times explains the corresponding ravines and valleys, which are formed during such events.

We shall focus our attention only on those island masses that are part of the territory of the Russian settlements. We will start with the southern point of Prince of Wales Island (54°40′ N. Lat.) and the [King] George the Third Archipelago in Norfolk [Sitka] Sound, also called the Kolosh Archipelago, near Sitka Island, and its vicinity.[14]

The *Island Sitka* or *Baranof* [Baranof Island] extends from north northwest to south southeast, from 56°10′ to 57°38′ N. Lat., and 134°20′ to 125°26′ W. Long. On its eastern side this island is separated from Admiralty and Kuiu Islands by Khutsnou Canal or Chatham Strait, the southern entrance to which is called Christian's Sound. Toward the north the island Chichagof lies before Sitka (divided by Pogibshii Sound [Peril Strait]). Westward, among other smaller islands, is Edgecumbe or Kruzof Island. Between there and Sitka Island are the islands Partofshikof, Krestovskii [Krestof], and Iablonnyi [Middle?]. Farther southward is Sitka Sound[15] with comparatively few large islands.

The entire western coast of Sitka is carved out by many bays. One of the largest of these is Sitka or Serebrennikof's Bay [Silver Bay], which reaches southeastward 14 miles into the land along the latitude of the southern rim of Edgecumbe [Kruzof] Island. In the northern part of its opening lies the main factory, Novo-Arkhangel'sk (57°03′ N. Lat., 135°18′ W. Long.). Parallel to the Bay of Sitka a second bay cuts in southeasterly. Then a third cut-in follows, behind which extends the so-called Deep Lake (Glubokoe) [Redoubt Lake], which is bottomless according to legend. This lake empties into the bay with an extremely swift confluence, similar to rapids. It forms cataracts, on the sides of which Fort Ozerskoi was erected, as well as a water mill (Shmakof's Mill). The lake just mentioned lies eight feet above the level of the bay. It is one Italian mile wide and eight miles long, and it receives Beaver River (Bobrovaia Reka) from the northeast. Near the southwest end of this lake there are hot springs (56°51′ N. Lat., 155°19′ W. Long.). Between these springs and the Ocean fort are (at 56°53′ N. Lat., 135°15′ W. Long.) Kliuchevskaia Mountain (Spring Mountain) and, close to the redoubt itself, Dranishnikof Mountain (57°54′ N. Lat., 135°15′ W. Long.). The four-sided Pyramid

Mountain (57°58′ N. Lat., 135°17′ W. Long.) lies close to the ocean between the above-indicated second and third cutouts.

The coast of Serebrennikov Bay seems to fall off with particular abruptness. There enters into its innermost eastern corner a little river with a waterfall, 50 fathoms high. And farther northward, Kupolnaia Gora [Cupola Peak] (57°00′30″ N. Lat., 135°06′ W. Long.) and Kamennaia Gora (57°02′ N. Lat., 135°09′ W. Long.) come close to the shoreline. North of the Fortress Novo-Arkhangel'sk [New Archangel] lies Harbor Mountain [Harbor Peak] (Gavanskaia Gora, 57°05′ N. Lat., 135°19′ W. Long.) with about 800 ft. elevation. At its base is another sweet-water lake. To the east, Verst Mountain [Mount Verstovia] (Verstovaia, 57°03′30″ N. Lat., 135°14′ W. Long.) is the highest on the island, with two peaks which are divided by a small ravine. According to Hofmann's barometric measurments, the higher of the two reaches 3,152 ft. Par. [Parisian] and is pointed. The other one is more rounded and has an elevation of only 2,376 ft. Par. The Kolosh [Indian] River, which has its source west of there, flows along the base of this mountain into the bay.

At the entrance of Sitka Sound the eye beholds an impressive view of this wild, picturesque region. From close by the seashore, high, cone-shaped mountains arise on all sides, with steep slopes and deep ravines. To the left is the extinct volcano Mt. Edgecumbe, 2,676 ft. [English] high. In front of the observer there is a group of rugged islands, in the back of the bay, the fortress, and elevated on a bare rock, the residence of the commander.[16] The wild character of the Natives, and the inhospitable nature of the land, consisting of rocks and permeated with bogs and thick forests, are the reasons why geologic investigations on Sitka have, up to now, had such little success. Thus in Postels's investigation during Lütke's voyage, we find reflected in the geologic part on Sitka only Hofmann's investigations, without that author being mentioned. In the deliberations which follow here, we shall concern ourselves primarily with E. Hofmann's observations.[17]

In the region of New Archangel the rocks consist of fine-grained, flinty graywacke. This formation contains longer and shorter strips of clay-shale. Sometimes this comes in layers intermingled with the rock. The commander's residence is located on a butte which, unlike most of the others, is not overgrown. It consists of black clay-slate. The graywacke changes from a conglomerate to quartz, lydian rock, and sienna with an adhesive of gravel material, and a fine-grained, almost crystalline rock with a binder of clay. According to Postels, this material has many fissures, the inner walls of which are lined with iron oxide. Therefore, the surface of the rock appears yellow, brown, or red in color.

A cave ten fathoms high is located a little south of the entrance to Sitka or Serebrennikov Bay and is probably embedded in this kind of rock. Hofmann searched for the location where the clay-slate changes to graywacke, near the fortress and the spit of land where the windmill is located. He found the place of changeover southwest towards south, and south-southwest under 75°.

Verstovaia or Verst Mountain consists primarily of graywacke, which forms a conglomerate in the smaller of the two peaks. On the higher, pointed summit, however, it is more finely grained and intermingled with clay-slate, which probably makes up the point as well. The shores of the bay where the Shmakof Mill is located, are lower but steeper than those near New Archangel. They, too, consist of strata of graywacke, which alternate with layers of clay-slate.

Both shores of Deep Lake [Redoubt Lake] are high and steep. The one on the right, or northwestern side, is overgrown with stands of conifers. The left bank is composed throughout of beautiful syenic granite of white coloration. The right bank is covered from the bottom upward, first with shaly graywacke, then with 2 to 3 fathoms of an imperfectly stratified mass of graywacke, which changes into porphyry, layered in separate plates. Then follows a finely grained graywacke. These layers seem to fall in from the northwest. Thus the length of the valley of the lake lies between syenic granite and graywacke.

On the strip of land which divides the sea from the southwest end of Deep Lake, syenic granite and graywacke border on each other along the tangent line of the right bank of the lake. Thus, graywacke expands toward the northwest, syenite toward the southeast. The latter contains nests and veins of graywacke in this location. And from a chasm in these rocks, which rise 200 ft. above sea level, three hot springs of 53° Reaumur mean temperature rise to the surface through fissures on the side facing the sea. There is evidence that these healing springs contain Ca, C, and S. The latter element separates in large quantities, coating, after a short while, the troughs and containers installed in the place with a thick crust of sulfur.[18] According to an analysis, which Mining Engineer Ivanov prepared in 1842, 100 parts of this sulfur contain:

Sulfur 97% 1 through 8 total the other 3%
1. Calcium, 2. Iron, 3. Manganese, 4. Soda, 5. Chlorine,
6. Ammonia, 7. Silicic acid, 8. Alumina.

1, 2, 4, 5 and 6 appear in larger quantity than 3, 7, or 8 (*Transactions of the Mineralogical Society*, St. Petersburg, 1842, pp. cvii and cviii).

In addition to the clay-slate, which is supposed to make up Cross [Krestof] Mountain (on Krestovsky Island), as well as Pyramid Mountain [Mount Kinkaid], we received from Voznesenskii genuine roofing plate from Japonski Island, where the magnetic observatory is located, and also clay-slate, rich in quartz, which gradually changes into complete flint slate. Furthermore, there is hornblende stone (which is found close to the fortress). A piece similar to this

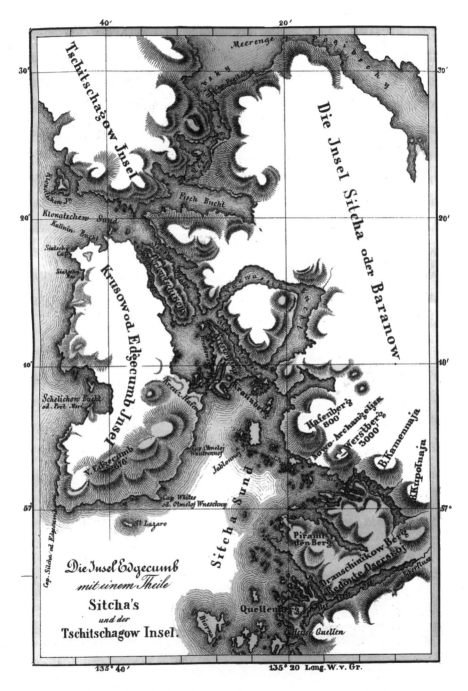

Map 2. Edgecumbe Island with parts of Sitka and Chichagof Islands.

sample, containing inserts of magnetic iron and iron gravel, is part of Hofmann's collection. There are, furthermore, strata of jasper, carnelian, well-preserved calcspar [crystalline calcite] and double spar [iceland spar], siderite and hard coal,[19] and finally a red bolus, which the Natives chew.

Through Fischer's efforts we have come into possession of a large amount of jasper, a large piece of suet shale [?], and a boulder of limestone with *Catenipora escharoides*.

In Kashevarov's collection we have iron silicate, bituminous coal, and tree trunks (conifers) that have metamorphosed into ferrous claystone. And in Nakwasina Bay on the west coast of Sitka (opposite Krestovskii), Mining Engineer Doroshin found limestone, and, furthermore, silver and iron ore, a large enough quantity to be mined. More detailed information is lacking.

Edgecumbe, or *Kruzof Island*, was named after the dormant St. Lazarus or Edgecumbe Volcano, which rises close to the southwest coast (57°03' N. Lat., 135°40' W. Long.) and has an elevation of 2,852.2 ft. Par., which equals 2,676 ft. Engl. The Spaniards, Juan d'Ayala, etc., made it known in 1775 by the name San Jacinto. Cook saw this mountain on his third voyage (Vol. II, p. 66, German edition), on the first of May 1778. He observed no volcanic phenomenon, but found it covered with snow. In 1796 (according to Hofmann) the mountain erupted with fire and columns of smoke.[20] But in 1804, when Captain Lisianskii (who named the island after Admiral Kruse, *Kruzof*) scaled the mountain, he found it at rest.

At the foot of the mountain, close to the coast, there is (according to Hofmann) a porous basalt (or basalt-like lava), which surrounds glasslike feldspar and a small amount of olivine. This rock formation seems to make up the largest part of the coast. On a promontory of the southwest coast, vertical walls of a dense mass of basalt with high olivine content jut upward. One of these walls forms a flat arc on top, which is composed of vertical plates of basalt standing together like bricks in a vault. The inner space is filled up with dense rock. Other specimens reveal that this basalt changes into a less hardened basaltwacke or basalt-like dolerite, in the bubble spaces of which there are deposits of stilbite. This rock exists in small particles as well, and gives to the rock a white-sprinkled appearance. On the sides of these pieces there is pitch stone, which also occurs on a little-altered piece of clay-slate. From the east side of Mt. Edgecumbe, we received porous andesitic lava, with predominant albite and clearly distinguishable particles of olivine. The ingredients indicate trachydoleritic characteristics. There also is found a porous mixture of albite, or (for lack of clearly distinguishable twin-formation) glassy feldspar with pitch stone, which resembles porphyric slate wherever it densifies.

The way to Mt. Edgecumbe leads at first through a conifer forest with low, mossy soil. Then it rises steeply for about 100 ft. Thereafter, it traverses a flatland, which is the bottom ledge of the mountains. After several more rises in elevation, there follows a second, wooded high plateau, above which rises the isolated cone of Mt. Edgecumbe. Toward the northeast two hills or mountains can be seen, from where a chain of hills extends toward the sea, as it gets progressively lower in elevation. Boulders of basalt, wacke, porphyric slate, and pumice stone lie about everywhere.

The cone itself consists of clay-slate with pockets and veins of pitch stone. Its sides are covered partially with vegetation, and partially with ruins of pumice stone. Near the top of the mountain is the crater. Lisianskii says it is 4 versts in circumference and has a depth of 40 fathoms. The rim and walls of the same consist also of clay-slate with pockets and veins of pitch stone. This is the only crater of the volcanos of this region, the dimensions of which are familiar to us. The circumference of the same, 4 versts or 2,000 *Faden*, results in a diameter of 636 *Faden* or 44,152 ft. Thus it is twice as wide as the crater of the equally tall Mt. Stromboli, and of the five-times-higher Kliuchevskaia Sopka, and similar to that of Mt. Pinchincha. According to Lisianskii (Vol. II, pp. 124–126), the water of this crater drains out completely at times. Thus, in July of 1805, the crater was completely dry. In July of 1824, the bottom of it was covered with snow. And in the same month of 1827, there was a deep hollow or lake, filled with water (Postels). Captain Belcher, who saw Mt. Edgecumbe on the eleventh of September 1837, says about it (Vol. I, p. 92): "(It is) a high dome-shaped peak, on which streaks of snow and bright lines of reddish-yellow clay radiate from its apex . . . the bluff termination of its western slope is Cape Edgecumbe." Simpson (Vol. II, p. 175) claims (in April 1842) that it wears "a diadem of snow" almost the entire year. And he thinks that its activity has not ceased entirely, because in its surroundings there still occur earthquakes, hot springs, and occasional explosions of smoke, fire, and ashes. The latter information, however, could well be erroneous, since it is not repeated anywhere else.

On the left, or western, side of Krestovskii Channel there is, according to Hofmann, an outcropping of secondary rock at the coast of the island, composed of graywacke and porphyry. Here, too, a rock wall with steep sides runs into the sea. It consists of brown and black pitch stone porphyry, which encloses rounded masses of dense basalt. Volcanic rock on the island extends as far north as Sukhoi [Dry] Strait (Proliv Sukhoi), which as its name indicates, is at times completely dried out. On Partofshikof Island, the graywacke is found again. Krestof Island, as well as the banks of Olga Canal, consist of flinty graywacke, such as occurs near New Archangel. A small island in front of the exit of Olga Canal also consists of graywacke.

Chichagof Island

Just as Edgecumbe Island is divided from Partofshikof Island by the so-called dry thoroughfare, so Sitka is divided from Chichagof (formerly Iacobi) Island by the still-navigable strait Pogibshii Zaliv [Peril Strait]. It is at first narrow with dangerous places, where cliffs and cataracts occur. But then it widens out from Cape Pogibshii [Pogibshi Point] to Chatham Strait, the banks of which, according to Hofmann, consist of graywacke on both sides. Boulders of coarse-grained greenstone lie about, which contain magnetic gravel and iron sulfide. According to Voznesenskii, there are outcroppings of clay-slate, amphibole, and serpentine stone on the east coast of Chichagof Island in Spasski Harbor (on the west side of Chatham Strait). At the middle harbor there are rocks of jasper, and on Chudsnof Bay, on the left side of the entrance, there is diorite. A small islet within this bay has white to grey limestone along its shores, stratified horizontally in some places, and steeply upright in others. Its structure is thinly slated and wavy-patterned. It contains pyrite and changes gradually into sandstone, of which material the island's interior is composed.

CHAPTER TWO

✦ ✦

The Mainland in the Latitude of Sitka, Chugach Peninsula, Kenai Bay and the Alaska Peninsula

OPPOSITE [KING] GEORGE THE THIRD ARCHIPELAGO, the Stikine River flows into Prince Frederick Sound at 56°45′ N. Lat., and 132° W. Long. We received from Simpson (Vol. I, p. 230) the following report about this coast. "[W]e weighed anchor (from Sitka) about five in the morning of the 30th of September (1841), but were obliged to bring up for the night about half past three in the afternoon in Lindenberg's Harbour (Frederick Sound). In the morning…the weather was cold and squally, while a little snow, that fell in the night, had partially whitened the green ice that filled the ravines of the mountains; and the channels were traversed by many restless masses which had broken off from the glaciers. In short, nothing could exceed the dreariness of this inhospitable coast." A redoubt had been constructed on the Stikine River. In its environs, Voznesenskii found mica schist, which contained a large number of garnets (Granatoeder–Leucitoeder, p. 202). They also are found in the riverbed. The Stikine River probably cuts through that mountain range which runs parallel to the seacoast up to 60° N. Lat., turning due north from there. It could be perceived as part of the Columbia Range, which as we saw previously, presses close to the coast opposite Quadra Island, from which point onward it becomes part of the composition of the coastal range. The mountains of Sitka, Crillon and Fairweather, would in that case form part of a chain which connects with the California Coast Range. This chain is frequently interrupted by the sea,[21] and northward, a short distance before Mt. St. Elias, it changes its main direction.

This assumption is justified, furthermore, because the rugged appearance and the wealth of islands of this coast become less so beginning with Cape Spencer (58°13′ N. Lat.), or Cross Sound. The region about Cape Spencer (perhaps a branch of Crillon Mountain) apparently corresponds with Cape Flattery to the north, and with the structure of the land toward the south. Only Port Francais (Lituya Bay) near Mt. Crillon has the characteristic form of the Californian bay. The mountain ranges along the coast of the mainland lean close to the sea, with gigantic precipices and glaciers. Their summits are little removed from the seacoast and have elevations of 8,000 to 9,000 ft., while their highest points (Mt.

St. Elias, Mt. Fairweather, and Mt. Crillon) reach an elevation usually seen only in the Andes. Mt. Crillon, at 58°45′ N. Lat. and 137° W. Long., is the first mentioned by La Perouse (Vol. II, p. 219). It has not been measured yet, but is said to be hardly less high than its neighbor. Mt. Fairweather (Vancouver, Vol. III, p. 204), Gutwetterberg, Cerro de Buen Tiempo, or Gora Khoroshie Pogody, lies at 59° N. Lat. and 137°30′ W. Long. According to different reports, its elevation varies between 13,819 ft. and 14,003 ft. Par. (14,708 ft. Engl., according to the map of the Hydrographic Department). Mt. St. Elias (cf. Bering's second voyage) in 60°17′ N. Lat. and 140°51′ W. Long., has an elevation of 16,758 ft. (Buch, *Canar. Inseln*, p. 390).

The entire coast from Cape Spencer to the mouth of the Copper River is, in spite of frequent visits and surveys of the same, extraordinarily little known with regard to its orographic condition. But even less so is the interior of the country known. Near Port Francais, Lamanon (*Voy. de La Pérouse*, Milet-Mureau II, p. 213) collected well-preserved petrified specimens of enormous size at a point 200 *Tois*. above sea level, specifically *Pecten jacobaus* (le manteau royal ou la coquille de St.-Jacques). The shores of this bay consist of 800- to 900-*Tois*.-high mountains. Only their peaks are covered with snow (July 1806). They seem to consist entirely of shale, which is beginning to deteriorate. Here was found: iron ocher, copper pyrites, garnets, schorl, graphite, bituminous coal, slate, hornstone, and granite.

On the whole, the mountains of this coast are composed of granite or shale, bare of any vegetation, and covered with perpetual snow. They rise abruptly from the water's surface, as they form a kind of quay. And they are so steep that mountain goats cannot climb them above 200–300 *Tois*. All the canyons are filled with enormous glaciers, the top of which cannot be seen, while the foot of them reaches into the sea. Soundings taken the distance of a tie-cable away from shore revealed that the bottom could not be found at 160 braces (*Brassen*) (La Pérouse, l.c.).

Vancouver's maps and data about the mountain range, which extends along the coast from Mt. Fairweather to Mt. St. Elias, have remained the most complete to this day. We will quote his description and his conception of the glaciers (Vancouver, French edition, Tome III, p. 233, from July 6, 1794 [Eng. ed., London, 1790, vol. 3, pp. 209–210]), which has recently been commented on by Belcher (cf. p. 15):

The base of this lofty range of mountains now gradually approached the sea side; and to the southward of cape Fairweather, it may be said to be washed by the ocean; the interruption in the summit of these very elevated mountains mentioned by Captain Cook, was likewise conspicuously evident to us as we sailed along the coast this day, and looked like a plain composed of a solid mass of ice or frozen snow, inclining gradually towards the low border; which from the smoothness, uniformity, and clean appearance of its surface, conveyed the idea of extensive waters having once existed beyond the then limits of our view, which had passed over this depressed part of the mountains, until their progress had been stopped by the severity of the climate, and that by the accumulation of succeeding snow, freezing on this body of ice, a barrier had become formed, that had prevented such waters from flowing into the sea. This is not the only place where we had noticed the like appearances; since passing the icy bay mentioned on the 28th of June [*sic*], other valleys had been seen strongly resembling this, but none were so extensive, nor was the surface of any of them so clean; most of them appearing to be very dirty. I do not however mean to assert, that these inclined planes of ice must have been formed by the passing of inland waters thus into the ocean, as the elevation of them, which must be many hundred yards above the level of the sea, and their having been doomed for ages to perpetual frost, operate much against this reasoning. . . .

Farther southward [north and west?], from Cape St. Elias [Point Manby?] to Cape Suckling, the coast and the two islands, Kayak and Wingham, near the last-named cape, seem to have undergone manifold changes since the time they became known. This is a point which has heretofore not been considered when attempts were made to ascertain the location where Bering had been closest to the mainland of America, or where he has touched on it. Let us consider what Captain Belcher says (*Voyage Round the World*, I, pp. 75–82):

All our transit bearings and other observations, plainly indicated the charts to be erroneous about this region. A river appears to flow near Cape Suckling, which has not been noticed.

Our attention was suddenly attracted by the very peculiar outline of ridge in profile, which one of our draughtsmen was sketching, apparently toothed. On examining it closely with a telescope, I found, that although the surface presented to the naked eye a comparatively even outline, that it was actually one mass of small four-sided truncated pyramids, resembling salt-water mud which has been exposed several days to the rays of a tropical sun, (as in tropical salt marshes,) or an immense collection of huts.

For some time we were lost in conjecture, probably from the dark ash colour. But our attention being drawn to nearer objects, and the sun lending his aid, we found the whole slope, from ridge to base, similarly composed; and as the rays played on those near the beach, the brilliant illumination distinctly showed them to be ice. We were divided between admiration and astonishment. What cause would produce those special forms? If one could fancy himself perched on an eminence, about five hundred feet above a city of snow-white pyramidal houses, with smoke-coloured flat roofs covering many square miles of surface, and rising ridge above ridge in steps, he might form some faint idea of this beautiful freak of nature.

Kaye's Island, viewed from the eastward, presents the appearance of two islands. The southern is a high table-rock, free from trees or vegetation, and of a whitish hue; the other is moderately high land for this region, with three bare peaks; its lower region being well-wooded.[22]

Wingham Island, which can be seen to nearly its whole length between Cape Suckling and Point Le Mesurier, (the north part of Kaye's Island,) is moderately elevated, rising in three hummocks, which are bare on their summits. The southern at a distance, owing to the lowness of the neck, appears separated. The whole is well clothed with trees.

In one direction from the southward, Cape Suckling exhibits on its bower profile, the brow, nose, and lips of a man. It is a low neck, stretching out from a mountainous isolated ridge, which terminates about three miles from it easterly, where the flats of the ice pyramids just alluded to terminate.[23] Apparently the river or opening near Cape Suckling flows round its base. There is little doubt but that we may attribute the current to this outlet, arising probably from the melting of the snow. We had less strength of current after passing this position. Immense piles of drift-wood were noticed on each side of the opening, but *none elsewhere*.

On the morning following [the fifth of September 1837] it was cloudy, with rain, and the breeze springing up compelled us to trip. Towards the evening it cleared up, and we were treated with a most splendid picture of St. Elias and all the neighbouring peaks, in full beauty, not a vapour near them. Each range is in itself an object worthy of the pencil, but with the stupendous, proud St. Elias towering above all, they dwindled into mere hillocks, or into a most splendid base on which to place his saintship.

Although Vancouver describes St. Elias as "in regions of eternal snow," yet his edges, to the very summit, present a few black wrinkles, and the depth of snow does not, even in the drifts, appear to be very deep.

My anxiety to reach Point Riou and obtain observations on it, induced me to hold on by the land. Indeed there was no other chance of overcoming the current. The coast presents so little to recognise in Vancouver's chart, that I despair of doing more than fixing the position of Mount St. Elias, which, if Kellett has been successful in seeing from Port Mulgrave, will be now secure.

Towards noon the breeze favoured us sufficiently to reach into Icy Bay, very aptly so named, as Vancouver's Point Riou must have dissolved, as well as the small island also mentioned, and on which I had long set my heart as one of my principal positions. At noon we tacked in ten fathoms, mud, having passed through a quantity of small ice, all of a soft nature. The whole of this bay, and the valley above it, was now found to be composed of (apparently) snow ice, about thirty feet in height at the water cliff, and probably based on a low muddy beach; the water for some distance in contact not even showing a ripple; which, it occurred to me, arose from being charged with floating vegetable matter, probably fine bark, &c.

The small bergs or reft masses of ice, forming the cliffy outlines of the bay, were veined and variegated by mud streaks like marble, and where they had been exposed to the sea, were excavated into arches, &c., similar to some of our chalk formations. The *base* of the point, named by Vancouver Point Riou, probably remains; but being free, for some distance, of the greater bergs, it presented only a low sand or muddy spit, with ragged dirty-coloured ice grounded. No island could be traced, and our interest was too deeply excited in seeking for it, to overlook such a desirable object.

On our inshore tack we had five fathoms and three quarters, and were therefore quite close enough to make certain of our remarks, short of actual contact, which the favourable breeze would not admit of without some more important results.

We edged along, keeping within a mile and a half of the shore, carrying from ten to fifteen fathoms, until night, when we bore away to cross Beering's Bay, and rejoined our consort in Port Mulgrave.

I perceive in Vancouver, (vol. iii. p.204,) twenty-three fathoms was his nearest approach, and within one league. He also terms it "low, well-wooded, with a small detached islet, a little to the westward." Also, "Eastward from the steep cliffs that terminate this bay, and from whence the ice descends into the sea." It is very probable there has been a misreading of his manuscript, or that severer weather had covered his trees with ice, for we saw none, and that portion of the coast was examined with his voyage constantly before me, and the discrepancies discussed with our spy-glasses on the objects.

Our observations and speculations, on the motion of the ice now before us, led us to suspect that the whole of the lower body is subject to slide, and that the whole of the substratum, as frequently found within the Arctic Circle, is a slippery mud. I am satisfied that this is the case in Icy Bay, as one berg, which was well up on the shore, moved off to seaward; grounding again near what I took for Point Riou.

This leads me back to our observations on the mathematical forms observed on the 3rd, after passing Bingham Island, and I perceive that Vancouver notices not only the ice, but (at p. 209, 210, vol. iii.) attempts to account for its formation, remarking that the ice observed (before reaching Point Riou and to the southward) was not so clean, "most of them appearing to be dirty." How came they so?

If the dark, "dirty" ice had been near the beach, it could readily be accounted for, by having been agitated with the beach mud, and forced up by gales. But the reverse is the fact. The darker ice was on the high ridges, and the bright near the sea. Only the theory of a slip would allow of its moving down the inclined plane without disturbing its mathematical arrangement. Vancouver's visit occurred in the latter end of June, ours in the early part of September.

In Icy Bay, the apparently descending ice from the mountains to the base was in irregular, broken masses, tumbling in confusion, similar to ice forced in upon the beach by gales of wind. They were doubtless detached masses from the mountains. But near Cape Suckling the inclination of the steps was very slight. . . .

From these not always exact descriptions we can discern that Captain Belcher has had the opportunity to observe glaciers and their movement, which were unfamiliar to him, and in this location most likely faster than elsewhere observed. A closer investigation of these would be of considerable interest. It seems on the whole, that glaciers have an extraordinary distribution[24] along this coast, and ought really be considered on par with the types of rocks. Even the most marginal of travelogues mention them, how much more so, then, in the following, more detailed report on the Copper River.

According to Klimovskii's Voyage, as edited by von Wrangell (Baer and Helmersen I, p. 162), the river called Copper River (Mednaia) along its upper run, and Atna River on its lower run, with its outflow at about 60°15' N. Lat. and 144°20' W. Long., comprises a separate drainage. They flow down from high mountain ranges which go in part (E) from Mt. St. Elias toward the NE, and otherwise (W), also in a NE direction across the innermost angle of Cook's Inlet (Kenai Bay). They divide the sources of those rivers which empty into the Bering Sea (watersheds of the Kuskokwim and Kvikhpak [Yukon]) from those which flow toward the Atna River that empties into the Pacific Ocean.

The Atna empties into the ocean with five arms; and at its delta it forms large sandbanks, which reach far into the sea. They extend mainly toward Nutchek [Hinchinbrook] Island (Khtagaluk). The river has broken through the Yakutat chain of mountains, which sweeps from Mt. St. Elias along the coast where the ravines are containers of permanent ice. The river undermines these masses of ice. As a result of this, large chunks are torn loose, which crash into the river with a gigantic report.

The mountain ravines, filled with ice 20 fathoms thick, are about 1.5 versts wide at the river. And in some places the ice is covered with dirt on top, where moss, berries, and alders grow. There are mid-river icebergs, covered with fresh vegetation and ripe berries. Above the rapids, which have developed where the river intermingles with the glaciers of the Yakutat Mountains, no more ice is found, and one enters a terrain which is no longer exposed to the wind and fog from the sea. These winds and fogs are limited to the coastal regions below the rapids, which the Ugalenzians [Russ. Ugalentsy; modern Eyak], too, inhabit only during the summer.

About 150 versts above the rapids, the very swift little river, Chechitna [Chitina?], flows into the Atna River. It has its source in a lake 150 versts east of this confluence. Now it is in the mountains along the banks of this river where in ground-elevations pure copper is found, in pieces weighing up to 1 *pud* (40 lbs. Russ.), but more often just a few pounds; also whetstone and mica. Every year, when the ice on the lake breaks up, the Chechitna River overflows its banks and floods the land at its mouth with such rapidity that the inhabitants are forced to flee quickly to the mountains to escape their ruin. One and one-half versts above this confluence on the banks of the Atna River, a hut with storage rooms (called Odinochka) has been built for a Russian who maintains a barter trade with the natives. Along the entire distance, from the Odinochka to the mouth of the Atna River, one observes mountains on both sides of the river. And the banks are rocky and overgrown with fir trees, poplars, land-willows, and birch trees.

Farther northward one comes upon fairly extensive flatlands between the mountains on both sides of the Atna River. On the left side, in view of the Odinochka, there rises a high, conical mountain.[25] It throws out fire continuously, and its peak is crowned with permanent snow. It is not connected to the mountain range, but stands alone. Several times each year this country is subject to violent earthquakes.

After receiving several small rivers and brooks from both sides, the river Atna divides into two main branches, as it seems, 1°48' north of the Odinochka. The right arm has its source in a big lake, Mantilbana [Tazlina?], a distance of five days' journey from its confluence with the left arm. That arm rushes along between the mountains with such speed that it cannot be navigated upstream.

Below the confluence of the Chechitna River, a small river flows into the Atna from the west. This river comes from a lake, where there is a comfortable portage to Chugach Bay [Prince William Sound]. In wintertime, the natives reach the northern part of Cook's Inlet, traveling straightway over swamps, lakes, and mountains. (Baer and H., *Beitr.*, l.c.).

We do not believe what one could be tempted to assume, namely, that several mountain ranges are sweeping between NE and NW, radiating outward from Mt. St. Elias into the land; with one of them along the coast, partly forming the banks of the Atna River, and rising up, on the other hand, as single mountains, volcanos, or high plateaus. If a mountain-hub is to be assumed to exist for the mountain ranges which stand against each other from SE and SW (over Alaska, Chugach, etc.), then the same should be sought farther inland, at 61° N. Lat., or perhaps 65° N. Lat., at the sources of the Copper River. It is, however, most probable, that the mountain ranges become ever lower, the farther they reach into the mainland, and that the latter will take on the characteristics of the flatlands of Northern Asia.

Between the Atna River and the Susitna River or Kenai Sound (Cook Inlet, Kenai Bay, or better Tlinaiskii Bay, according to Zagoskin, part I, p. 88), and Chugach Sound [Prince William Sound], lies the Chugach Peninsula [Kenai Peninsula], which at 61° N. Lat. is connected to the mainland by only a narrow strip of land. Two bays, which cut deep into the land, extend close together here. This formation, and the Kaknu [Kenai] River,[26] which flows from Lake Skilloch [Skilak], close to Voskresenskaia Bay [Resurrection Bay], with its estuary on the west side of the peninsula near Fort St. Nikolaus—this, as well as Chugach Bay farther to the south, give us a clear view of an island formation which had in this case not come about. Lake Mantilbana, too, and other lakes close to the coast, from where rivers flow into the land before falling into the Atna River, are typical of the peculiar character of this mountain region, and of the coast which arises with the Yakutat Mountains, and lies transversely before the rows of mountains that spread inland (cf. Vancouver's *Atlas* and Profile on Pl. XV, furthermore Map 5 of this essay).

The distribution of the mountain ranges, beginning with Mt. St. Elias, resembles the region about the estuary of the Fraser River, where similar events are manifested, only in the opposite direction. If we extend the analogy further, then we can assume behind the sources of the Atna River a mountain range following the major direction of the North American mountain ranges but perhaps more northward. A closer observation of the structure of the peninsulas, Chugach and Alaska, also teaches us that here two parallel coastal elevations can be pursued from SE to NW: one for the Chugach Peninsula and the islands Shuyak, Afognak, Kodiak, Trinity and Ukamok; the other for the Alaska Peninsula and the adjacent Aleutian Islands.

The former elevation continues northeastward from Chugatsk, with the mountain ranges between the Susitna River and the Copper River. The latter line up NE of the Alaska Peninsula with the Chigmit Mountains.

But it is yet to be firmly ascertained whether the Yakutat Mountains (Robin Mountains), which extend from Mt. St. Elias along the coast, are to be seen as a continuous range, together with the Chugach and the Truuli (Kenai) Mountains,

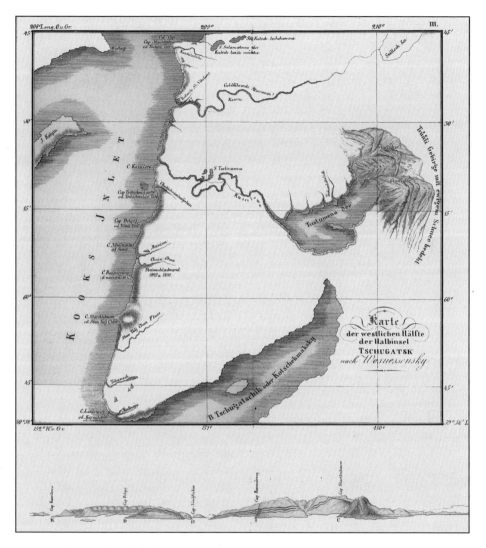

Map 3. The west half of Chugach [Kenai] Peninsula.

or whether they cease to be an indepedent range before they reach Chugach Bay
[Prince William Sound] (cf. Map II).

This is what von Wrangell says about the east coast of Kenai Sound (Baer
and H., *Beiträge*, Vol. I, pp. 168–179):

> The coast of Cook Inlet, which I have seen from Anchor Point Promontory
> (Anchor Cape, Mys Iakornoi, Laidennoi, or Kassnachin), to Fort Nikolaev
> (60°33′44″ N. Lat.) at the mouth of the little river Kaknu (Kenai River) con-
> sists of a steep ridge, about 100 feet high, with now and then some stands of fir
> trees. The mountains sweep eastward deep inland; and the high coast consists
> of clay and sand. The clay is of a bluish color and lies below. On top of it is a
> twenty-foot-high layer of sand. From the blunted Point Kassilov (Kasilof)
> southward, and in Chugachik or Kochekmak (Kachemak) Bay itself, bitumi-
> nous coal is found in several places in the middle of the steep bank, in all the
> stages of transition from bituminous wood to glance coal [pitch coal], in hori-
> zontal deposits of a thickness of one *arshin* or above.[27] Along this stretch one
> encounters no permanent rock formation. Those rocks which are scattered
> along shore and, in several other places of the bay, partially hidden under water
> and partially protruding above the water's surface, are enormous boulders of
> white granite with large crystals of feldspar.

Mr. Voznesenskii has submitted a report on the findings of bituminous coal
on the right bank of Cook Inlet, from which we excerpt the following commentary, as we add a map and profile (No. III).

A little north of Cape Starichkof (Stuk-talj-chin)—S of Ninilchik, 150 paces
from a waterfall, not very rich in water, but distinguished by a deep and wide
basin—there are two parallel seams of bituminous coal, 10 *vershok* thick, within
a loose lime marl. They are exposed 16 fathoms above the *laida* (the places lying
dry during ebb tide), and 6 to 10 fathoms below the upper crest of the bank. Both
seams, to which soon a third one is added, extend NNE almost to the first protrusion of Cape Ninilchik. They disappear at the low coast near the mouth of the
Chnik Chak River, close to the place where a subterranean bituminous coal fire
occurred in 1829 and 1830. The thickness of the individual deposits diminishes
from the top downward. A layer of fine yellow sand, six *arshin* thick, forms the
floor of the lowest one. Then follow 3–4 *arshin* of sand and lime-marl with clay
intermingled to the next seam. And between the middle and the upper layer of
coal, there are 3 fathoms of the same material.

On the second protrusion of Cape Ninilchik it can be observed how the coal
seams emerge again from the ground, rising with inverse folds SSW [*sic*] up the
shoreline. At first the coal seams have the earlier thickness. But they trace no longer

straight, but wavy lines; and they disappear progressively the farther northward they are observed. At the 5-fathom-high bank of Cape Kukistan or Kolgoi, only narrow, ribbon-like seams of coal can be seen. Their base is sand and grey clay and the layers in between consist of white limestone. Above the coal there is a layer of vein-like limestone, and then a layer of peat (tundra layer). At Cape Chikhkalansk the coal disappears completely. Added to the description of the hollow-like stratification of coal seams in Voznesenskii's dispatch, is the deepest layer of loam, or rather clay to marl, which changes into clay-slate to coal-slate, or better: layers of the latter, which contain bituminous coal to anthracite. Among the plant impressions which are found in this coal-slate, there can be discerned leaves of *Alnus*, and a species of *Taxodium*, which, however, is not found alive in this place. That 20-foot-thick layer of sand, mentioned by Wrangell, is represented in Voznesenskii's report by a brick-colored, clay-like sandstone. Then we find a conglomerate of clay-slate (lydian stone) with rounded flint and flint-clay as a bond. Finally, there is a large amount of more or less preserved tree trunks, crystallized and changed into brown iron, and large sections of peat to brown coal.

Boulders are composed as follows: granite with a predominance of quartz, and small amounts of feldspar, little scales of mica, and finely interspersed garnet; amphibolite of dark-green, leafy-radiating hornblende with a little quartz and a trace of pistacite; soapstone-shale; red boulders of porphyry; quartz-porphyry. And from Kalgin Island, amphibolite.

Along the western side of the bay, according to Wrangell, there is a chain of high mountains,[28] with several summits or peaks (sopki) among them which are covered with permanent snow. The highest of these mountain tops, visible from Nikolaev Redoubt, is the one which Cook had recognized as a volcano.[29] Its pointed top emitted smoke continuously. The sides of the mountain are interspersed with deep fissures, which can be clearly discerned from the Redoubt with the naked eye—a distance of 133 versts. The entire visible part of this mountain, here known as the fire-belching peak Ilaeman (Iliamna Volcano) is wrapped in permanent snow, from where on sunny days it appears so brightly gleaming, that even a trained eye would estimate its distance much diminished. According to the true compass, its location was determined to be 62°11′30″ SW of the Redoubt, and its elevation 12,006 ft., that is, 1,085 ft. higher than Mt. Aetna. On the map of the Hydrographic Department, Mt. Iliamna is located, according to Cook, at 60° N. Lat. and 153°15′ W. Long. Its crater, as Cook indicates, is on that side which faces the Inlet. And its location is not far below the peak. Those voyagers (at the end of May 1778) did not take much notice of it, because only some white smoke could be seen, but no fire.

Opposite the Redoubt, there arises another mountain peak, at 152°45′ W. Long. and 60°30′ N. Lat., which is about 11,270 ft. high (Engl.). It is found indicated on the map of the Hydrographic Department by the name of *Gora Vysokaia* (High Mountain), as a smoking mountain [probably Redoubt Volcano]. It was first made known by promyshlenniks [early Russian fur traders], and the natives call it *Uyakushatch.*

The southern coast of the Alaska Peninsula was surveyed by Lieutenant Vasil'ev and Voronkovskii. The former did his survey in 1832 from Cape Douglas (at the mouth of Kenai Sound) to Cape Kumtiuk [Kumliun],[30] and from this point onward Voronkovskii surveyed (1836) to the Chitkuk Promontory.

The peculiar character of the Alaska Peninsula[31] (Aliaska, Aliaksa, Alaeksa and in Aleut: Alakh–khak) justifies that we consider it, as well as the adjacent groups of islands, by themselves.

Alaska Peninsula extends in the form of an arc from NE ¼ E to SW ¼ W, as part of a circle which could be drawn along the Aleutian Islands around an imaginary center at 65°30' N. Lat. and 177°45' W. Long. This circle would go from Bering Island, approximately across Cape Stolbovoi, Bear Island off the mouth of the Kolyma River, and Cape Barrow; so that the latter would form a line with the Stolbovoi Promontory, and the NE border of the Alaska Peninsula would fall into line with Bear Island. On the mainland a natural NE border is formed for the peninsula by the 50-verst-long Kamishak Bay[32] at the entrance to Cook Inlet between 59° and 60° N. Lat., in that this bay approaches the large Lake Iliamna, or Shelikhov Lake, from where the Kvichak River flows into Bristol Bay, or Kvichak Bay. In the southeast, Alaska Peninsula is bordered by the Great Ocean, from Cape Douglas (Kukhat in Aleut) to Isannakh or Isanotskii Sound [Bechevin Bay], with outlying islands and straits (such as Shelikof Strait). Northwesterly it is washed by the Bering Sea, from Isanotski Strait to Bristol Bay (59° N. Lat.). The greatest width of the Peninsula is 110 nautical miles near Naknek River, and its length from 153°30' to 163°–164° W. Long., is 450 nautical miles, according to Lütke. The southern coast is substantially different from the northern coast. The former has twenty sizable bays. These and the outlying groups of islands, the Eduokedokin [Semidi], Shumagin, Semenof [Simeonof I. ?], Pavlof, and Belkof Islands, give the coast a very rugged appearance. Along the northwestern side, on the other hand, there are only six larger bays and few islands (cf. the profiles in Lütke's *p. n.* No. 1–7).

Most narratives on Alaska usually mention the following: the acute and tall chain of mountains, not more than six geographic miles wide, is frequently interrupted or cleft by valleys. This opinion, however, still needs verification. This does not seem to be a single mountain ridge, but rather a *faîte géométrique,* or only a row of individual mountains of more or less high elevation, and of a volcanic nature (eruption chains), along a certain geographic line. They seem to exist as a continuation of the Aleutian mountain islands. Between these mountains, the ground is not much altered, so far as topography and composition are concerned. It forms so-called *perenosy*, portage places, which cannot at all be taken for mountain passes on the basis of available descriptions. Alaska offers us a clear view of island formations which have not come about. (The same has already been mentioned above with regard to the Chugach [Kenai] Peninsula, the SE coast of which is in like manner different from its NW coast.) A continuous mountain range can hardly be spoken of. It is possible that in earlier times, when the ground level was lower, today's Alaska Peninsula was divided into several islands, and that this condition was subsequently altered by a later rising of the land.

Let us now turn to a more detailed investigation[33] of the peninsula, so far as that is possible with the material at hand.

Behind the isthmus or portage-place (*perenos*), which is formed between Lake Iliamna and Kamyshatskaia Bay [Kamishak Bay], there follows a similar indentation between the Naknek River, the lakes where that river originates, and the valleys a man must cross to get to the southern coast.

A third lowland, or the edge of a new elevation is formed by the Uchaguk River [Ugashik or possibly Egegik], by the lakes Nugashik and Ninuan-Tugat [Ugashik Lakes, and Becharof Lake], and by Puale Bay, and Kialakvit Bay [Wide Bay] with the two rivers that flow into its western and eastern corners.

The fourth lowland is situated between Heyden Bay [Port Heiden] and Kizhulik [Kujulik] or Chignik Bay.

The fifth lowland or isthmus reaches from Cape Rozhnof to Pavlovskaia [Pavlof] Bay, or from Moller Bay [Port Moller] to Stepovoi [Stepovak] Bay, which are connected by two lakes and a river that flows into the latter. They form the eastern border of the District of Unalaska.

The sixth depression is situated between Isenbeck Bay [Izembek Lagoon] and Morozovskaia Bay [Cold Bay]. Moreover, it seems that the former is connected to Morzhovoi Bay by a low flatland and several lakes along the northwest coast.

If we imagine these depressions and a few lateral valleys inundated by the ocean, then we obtain islands along the dotted lines on the map. Their contours would by and large be comparable to those of Unimak. On the basis of these lines we can most easily obtain an impression of the area's geography.

Off the southern end of Kamyshatskaia Bay, at 59° N. Lat. and 152°52′ W. Long., there lies Shaw's Island (Shunakhtali in Aleut), followed by the far-extending Cape Douglas. Southwest of the latter a small distance away, is Four-Peak [Fourpeaked] Mountain (Chetyrekhglavaia, 58°45′ N.Lat., 153°30′ W.Long.).

Cook (III, v. 3, p. 384) saw smoke rising at the northern entrance to Shelikof Strait, in the proximity of Cape Douglas. He called the location Smoky Bay. It is not clear whether the smoke arose from the dwellings of the inhabitants, or perhaps from Four-Peak Mountain. Bering's Mt. Dolmat seems to have been Four-Peak Mountain, rather than Mt. Iliamna.

Lütke figures 150 Italian miles from Cape Douglas to Puale Gulf [Puale Bay], over which distance six bays cut into the land: 1) Swikshak Bay. Several small rivers flow into it. Its southern limit is Cape Nukshak. 2) The large Kukak Gulf [Kukak Bay] with Cape Igyak [Ugyak]. 3) The Bay of Kaflia, with Utikoi Island off its southern point; then Cape Atushagvik (58°05′ N. Lat.), and next to another smaller bay, 4) Kinakhagliuk Bay [Kinak and Amalik Bays], in front of which is Takhli [Takli] Island. Hereupon 5) Tagalak [Dakavak] Bay; and finally Katmai Bay, which ends with Cape Kubugakli. From here it is approximately 60 miles to Nunakalkhak Promontory (Nelyupiaki) [Cape Kekurnoi].

Along this entire stretch, the coast is alternately steep and flat. Information on higher mountains and mountain ranges is lacking. With Cape Atushagvik and Kulugmut Promontory the coastline comes closest to Kodiak Island. Shelikof Strait is a mere 21 miles wide at this point. But the islands Afognak and Shuyak, north of Kodiak, are farther away from the coast of the mainland.

On Katmai Bay, 2 miles from the sea, Katmaiskoe village lies in a flat, swampy region, next to a small river which flows into this bay north of this place. A plain extends northwestward from the village, which leads the way to the Naknek River. Jurassic deposits occur near this village. The ammonites sent by Voznesenskii, (*A. wosnessenskii*, *A. biplex*), and the belemnites (*B. paxillosus?*), and one *Unio* (*U. liassinus*), lead to that conclusion. A description of these is found in the appendix. Farther inland on the way to the Naknek River, *Tritonium anglicanum* (?) was collected from a tall rock wall, dangerous because of its avalanches. The same is also found in Tertiary layers on Unalaska.

In the proximity of Studenaia or Puale Bay, situated between Cape Nelyupiaki and Cape Aklek, there is the large expanse of Lake Ninuan-Tugat [Becharof], surrounded by high mountains which are cleft by a deep ravine at the bottom of the bay. On the whole, the peninsula seems to become more mountainous from here on. Kanatak Bay (between Cape Unalishakhvak and Igvak) is followed by the very wide Kialakvit Bay [Wide Bay]. Two rivers flow into its western and eastern corners. The background is mountainous, with Mt. Alai at 57°24′ N. Lat. and 156°15′ W. Long. The western coast is covered with glacial ice, which also occurs in the small Agripina Bay. This allows the assumption of the presence of significant elevations.

Adjacent to the last-named bay is Wrangell Bay [Port Wrangell] and harbor, before which lie the islands Davidof [David] and Poltava. Two rivers cascade into the inner bay. The shoreline is sandy. In the background toward the next bay, Mt. Chiginagak is visible (at 57°03′ N. Lat. and 156°45′ W. Long.).

Between the southern point of Wrangell Bay (Cape Providence, de la Providence, Provideniia, at 156° W. Long.), and Cape Kujugak [Kuyuyukak], there extends Chiginagak Bay. Thereafter follows a low coastline, cut in by two small bays (Nakkhalilek [Nakalilok] Bay and Amber Bay).[34] Then the coast becomes more mountainous again, at Aniakchak Bay. This bay begins with Cape Ayukhta [Ayutka] on the right, and ends on the south side with the island and promontory Kumlik. Near there is the large Sutwik Island. Between Cape Kumlik and Kumliun is located (at 56°32′02″ N. Lat.) Kizhulik [Kujulik] Bay; and in front of it Unavikshak Island. Southeast of Cape Kumliun lie the Semidi or Edokeev Islands. Ukamok or Chirikof Island is also considered a part of the group. Between Cape Kumliun and Cape Tuliumnit [Tuliumnit Point], there is Chigmit or Chignik Bay, which extends inland in the direction of Heyden Harbor [Port Heiden] on the opposite coast of Alaska. Nakchamik Island lies off Chigmit Bay. From here on, the coast juts out strongly the entire distance toward Cape Kupreanof (also called Ivanov, at 55°34′30″ N. Lat., 159°25′ W. Long.), or Stepovoi Bay [Stepovak]. There is only one deep indentation, Kupreanof Bay. A medium-sized river flows into it, which originates in a lake. Volcanic phenomena have been observed in the interior of the country along this distance. Off this coast there are the islands Chankliut, St. Mitrofan [Mitrofania], St. Paul, and St. Jacob. Stepovoi Bay cuts inland close to Moller Harbor [Port Moller]; and there begins the Shumagin Archipelago, the largest island of which is Unga. Opposite this island is Perenosnyi Bay [Balboa Bay] (so named because of a communication trail to Moller Bay). A good-sized river flows into this bay, at the sources of which hard coal is supposed to be found. On the eastern side of this bay there are hot springs (at 55°35′ N. Lat., 160°27′ W. Long.), also deposits with contents of petrified material (Veniaminov, I, p. 235). There follows Bobrovaia Bay (Beaver Bay), from where a *perenos* [portage] leads to the next bay. That bay, surrounded by brush-covered slopes, is Pavlovskaia Guba [Pavlof Bay] (Tatchik in Aleut), in front of which lie the Pavlof Islands. The bay has an opening of 12 miles. On both sides mountains rise toward the interior. Toward its head, however, the bay is shallow, filled with moss, and surrounded by several lakes which give it the appearance of a strait. Four small rivers flow into the bay.

Pavlof Volcano rises on the left side (westerly) at 55°24′ N. Lat., 161°45′ W. Long. It has two craters, the southernmost of which is active. But about sixty years ago (in 1786),[35] the northern one was supposed to have collapsed with a mighty detonation. Thereafter it has ceased to be active. According to Veniaminov (I, p. 222) there rises behind this mountain a jagged mountain ridge, which consists of many regular pillars and pyramids. Viewed from the SE, it looks like a castle or a monastery with towers.

From Pavlof Volcano, I. Voznesenskii sends us good bituminous coal[36] and clay-slate. And close to the Pavlof settlement, there is a clay deposit which becomes pulpy in wet years. In it are found peculiar rock-balls of clay-slate and clay-ironstone. When struck, they prove to be either compact or hollow; or they contain a round core (rattle stone, eagle-stone). This clay also contains a few fossils. On the east side of the bay there are strata of bituminous coal which seem to be worthy of mining, according to Voznesenskii.

Near Pavlof Volcano, as well as in the vicinity of the village, Tertiary strata of the youngest kind protrude with contents of *Cardium decoratum*, *Venerupis petitii* var., *Pectunculus kaschewarowi*, *Mya crassa, truncata, arenaria*, *Mytilus* (?), *Ostrea longirostris*, *Pecten*, and *Crassatella*. The description of these is found in Appendix I.

Southeast of the volcano is a plain that inclines gently toward the sea. But southward and southwesterly the coast is mountainous, and toward the east it forms an elevated plateau, according to Veniaminov. Another bay on the right almost extends as far as Bobrovaia Bay. Three small rivers flow into this bay.

From Pavlof Bay it is easy to travel to the north side of the peninsula by way of a mossy plain. The way leads across an isthmus of 1.5 versts' width to a lake, which in turn is a mere 150 *Tois.* distant from another lake. From the latter lake, a river of considerable size falls into a bay of the northern coast. This bay begins at Cape Rozhnof and merges with Moller Harbor. From east to west, this is the narrowest part of the Alaska Peninsula. It is completely covered with lakes. South of Pavlof Bay there follows Medve[d]nikovskii Bay [Bear Bay], set apart by only a small promontory, which branches out from Pavlof Volcano. In front of the bay is Dolgoi Island, which belongs to the Pavlof group of islands. Four small rivers flow into this bay, from which point there is an overland connection to the next, Belkovskaia Bay [Belkofski Bay]. On the western side of Bear Bay, a steep cape slopes into the sea, from where Medvednikovskii Volcano arises at 55°05' N. Lat., 162° W. Long. It presents the spectre of a ruined mountain that has collapsed in on itself.[37] From here a row of rock cliffs, hidden under water, extends all the way to Ilaeshek [Iliasik] Island. Farther inland a high plateau extends from the volcano. From this plateau there rises a mountain of considerable size. Toward the ocean a high plateau extends with a promontory beyond which the coast makes a sudden turn to NNE, forming Belkofski Bay. This bay is surrounded by mountains, which are cut through by several deep ravines as they descend toward the sea. From its east side there leads a connecting trail to the last-named bay, and from its western side one to the next following bay. Near the west cape of the bay (Mys Stolb, Pillar Cape [Bold Cape]), the rocks drop off abruptly into the sea. Forty versts south of Belkofski Village, there is a tall rock, surrounded by a great number of small rocks. It has been called the Sivuchii (sea lion) rock, because these animals are being killed here in great numbers. This west cape of the bay belongs to a fairly high mountain ridge which forms the division for the next following Morzhovoi Bay [*sic*—Morozov or Cold Bay].

This bay cuts deeply northward and approaches the large Izenbeck Bay [Izembek Lagoon]. Both of these large bays almost cut the whole peninsula in two. Only a narrow strip of land covered with gravel and sand divides them. This, according to Lütke, is the third east-west section through the mountain range that runs the length of the peninsula. On the west side of Morzhovoi Bay two small rivers flow into the sea. One of these drains three lakes, while the other one comes from the mountains. This western side is full of high mountains. Among them is the volcanic Morzhovoi Peak [North Walrus Peak?], at 55°05′ N. Lat., 162°37′ W. Long., which some time ago spewed flames, and smoked mightily.

A small island of high elevation lies off the western cape of this bay. But farther southward Reindeer Island (Olenii) [Deer Island] lies in front of Morzhovoi Bay. In the bay, opposite this island, there is a low promontory, composed of clay and rock debris, which rises gradually and noticeably from year to year.

Morzhovskii [Morzhovoi] Bay (also Morozhovaia or Walrus Bay) is not far from Izembek Lagoon. To the west is the mouth of a small stream, which originates from a big lake that almost fills the entire lowland in the north. Only an isthmus, 100 *Tois.* wide, divides this lake from the sea. It lies so low, that at high flood tide it is partly covered with water. West of Morzhovoi Bay, the mountain range rises again and continues without interruption to the SW end of the Peninsula. At the mouth of the bay, more towards the eastern side, are two small islands. On one of them (Amagat?), 54°54′ N. Lat., 162°50′ W. Long.) there is a hot spring. Khudiakov calls the east cape of this bay Animachichko [Cape Tachilni], and that of the western side Kakhidaguk [Kabuch Point?]. On the latter promontory, the first one east of the Strait of Isannakh [Isanotski Strait], there appears 50 *Tois.* above sea level a horizontal layer of petrified shells. On top of this lie horizontal layers of sand and clay, another 50 *Tois.* thick. Lütke (*p.n.*, p. 272) referred by mistake to a more easterly located cape (on the western side of Morozovskaia Bay) as the site of this phenomenon (cf. Veniaminov, I, p. 222, p. 236). According to Voznesenskii's specimen, these strata belong to the most recent Tertiary formation, as becomes apparent from the description of fossils in Appendix I. These are: *Mya crassa*, *truncata*, *arenaria*, *Tellina edentula*, and perhaps also *Tellina lutea*. The southern coast is extremely steep all the way to the Strait of Isannakh (Aleut for "hole") [Isanotski Strait]. Only close to Morzhovskoe village does the coast get lower.

Among Voznesenskii's specimens sent from this place we find coarse pyrite and, in larger quantities, graphite.

The region surrounding the village just named is low-lying, swampy, and covered with lakes. The village is located on the eastern side of the northern

Morzhovoi Bay, which ends on its western side with Krenitzin Cape. There are two coves within this bay: Protassof Cove, and the cove of the hot springs (Goriachikh Kliuchei). From Krenitzin Cape to Glasenapp Promontory the coast is sandy and not very high. Between the elevated Cape Glasenapp, with its cut ridge, at 55°14'08" N. Lat., 162°50'07" W. Long., a cut that makes it appear to be an island, and the low Moffet Promontory [Cape Moffet], there is the location of the large Isenbeck Gulf [Izembek Lagoon].[38]

The eastern side of this half is surrounded by high snow-covered mountains. Several islands lie in front of the bay, and a little farther out in the ocean, the island Amak, which is an extinct volcano. From Cape Moffet 50 miles upward to Cape Rozhnof, the coast is flat, interrupted only by two promontories (Leontovitch and Leiskof). The soil consists of peat. It is soggy, covered with moss, and abundant in lakes and good drinking water. Thereafter follows the earlier-mentioned Moller Bay [Port Moller], which extends 12 miles into the peninsula. Two lakes and a river connect it to Stepovoi [Stepovak] Bay.

Moller Bay has the only good harbor on the north coast of Alaska; and it is surrounded by mountains of medium elevation. Along this bay there occur strata which contain fossils. In front of it lies the east end of the flat island, Kritchkoi [Kritskoi] (Morzhovoi Island). The next noteworthy point is the fairly high and steep Cape Kutuzof. From here on, the coastline recedes northeasterly inland. Thirteen miles NE of Cape Kutuzof is the location of Cape Seniavin, at 56°23' N. Lat. and 160°02'07" W. Long. This cape is high and steep. Northeast of it there are three small mountains. The next bay, Graf Heyden, [Port Heiden], is little known. In 56°53' N. Lat., the low north cape, Strogonof, is a part of it. The bay is surrounded with peaty, mossy, wet soil, that seems to extend as far as the base of the fairly distant mountains.

In 57°05' N. Lat., emerge another two promontories; and in 57°10' N. Lat. a cone-shaped mountain can be observed. From here on, the mountains recede far from the coast and go farther inland.

There follows now, 8 miles from the mouth of the Sulima (Ugatchik) River [Ugashik], the high and sandy Cape Menshikof, which is surrounded on all sides by swampland. Thus, during high water the cape might become an island. From the north point of the Sulima River estuary to Cape Greig, and the river Ugaguk [Egegik], and farther onward toward Cape Chichagof, the coast is swampy and monotonous. But Greig Promontory is steep and a little more elevated than Cape Chichagof. The left (southern) bank of the Ugaguk River is very low. On the northern side there is flatland, on which 7 miles NE of the river mouth a small mountain can be observed. The Ugaguk River comes from Lake Ugashek [here, Becharof Lake is meant].

From Cape Chichagof to the Naknek River estuary the 30-mile-long, monotonous, swampy, flat coast is raised up like a wall, 100 *Tois.* inland. It is supposed to consist of sand and clay. There is a fairly large mountain at 58°35′ N. Lat.

The Naknek River has its source in the lake of the same name. At first it flows from east to west for a distance of about 50 versts; and then it flows, 1 verst wide, southwesterly into the sea. The flood tide rises here to a level of 30 ft. Its banks are steep; and they consist of a rich clay. Sixteen miles N ¼ NE from the mouth of the Naknek River there is, at 58°57′08″ N. Lat. and 156°54′ W. Long., the estuary of the Kvichak River, from which point, according to Lütke, the natural border of the peninsula can be said to begin.

The NW coast of the Alaska Peninsula,[39] as well as the sea bottom, especially opposite Amak Island, consists of ferruginous sand and volcanic rubble of rock with pumice stone; this according to Kastal'skii, the naturalist on the sloop *Moller.*[40] Among the strata close to Cape Rozhnof can be found chalcedony, syenite, common quartz, and petrified wood. But according to Veniaminov (I, pp. 26, 45 and 222) there also occur fossils at Moller Bay, at an elevation of 40 fathoms in a layer of very solid clay or clay-porphyry, the same as at Morzhovoi Bay.[41] Judging by some of the specimens, he seems to be describing a black, infusionless substance of flint. For the rest, this coast, too, is covered with ferruginous sand. And in the bed load, which contains lava, pumice and other kinds of rock changed by fire, there can also be found jasper, fine-grained granite, chalcedony, common quartz, porphyry and serpentine.

Hot springs are supposed to exist south of Moller Harbor (55°45′ N. Lat., 160°30′ W. Long.?). The same characteristics can be found in the region surrounding Cape Kutuzof.

In the region of the Sulima River (Ugashik), the appearance of the coast undergoes a change. The volcanic character is absent, and more granite can be found, with amphibole (syenite), jasper and quartz of different colors, likewise pumice stone and calcium deposits, which are obviously thrown out by the sea. Even fewer traces of volcanic activity are found at Naknek River, where there are deposits of grey granite, gneiss, coarse black amphibole, serpentine, porphyry and sandstone.

According to Postels (Lütke, Vol. 3, p. 27), the west coast of Alaska falls off steeply to the sea and at an elevation of 300 ft., it exposes parallel strata which are supposed to contain a large amount of bivalves. But it seems that this might indicate the deposits at Moller Bay, as indicated above. On one of the promontories there is an outcropping of a fine, firm sandstone. The natives use it as whetstone.

From what has been said, it appears that between the SE and the NW coast of the Alaska Peninsula, the same variations exist as they do along the Chugach Mountains. While the former coastline is rugged, featuring a large number of bays, and surrounded by many islands, rocks, cliffs and stone banks, both under and above water, the NW coast has few bays. It is shallow and monotonous. But it has a more healthful climate, no high plateaus, and a few good anchoring places.

Postels states that the Alaska Peninsula is traversed by three mountain ranges: Morzhovskii, Ivanov and Pavlov. But he says nothing further about their position and extent. Lütke does not mention these ranges at all. All he indicates is that "Parallel to its SE coast, a mountain range traverses the entire peninsula. Its southwestern end is high, where there are several snow-clad mountain peaks. Toward NE the elevation declines, and the range recedes from the coast the more the peninsula gains in width."

Postels's claims were not adopted by Veniaminov. Neither do they make sense. If, for instance, an Ivanov Range is supposed to exist, it should be situated near the cape of the same name (or Kupreanof Cape on the map). But it is easier to locate its position between the Morzhovskii and Pavlov Mountains. Therefore, we will not adhere to these names. Instead, we shall merely point out the individual, independently arising elevations.[42]

If we begin southwesterly with Isanotski Strait, we notice that from its steep, high coast, the mountains rise to a significant elevation toward the western side of Morzhovoi Bay. Regrettably, the highest snow-covered points, which are located more toward the eastern side of this stretch of land, have not been measured. Neither are volcanos indicated anywhere. Toward the north, between Morzhovskoe Village and the head of Morzhovoi Bay, the mountains fall off more gradually and not as abruptly as on the west and east coast. Then, along the distance toward the Bering Sea, there follow several lakes and an isthmus of 100 *Toisen* width, which at times is completely inundated. Perhaps that is why Postels calls this mountain range the Morzhovskii Mountains.

The eastern rim of Morzhovoi Bay and the western limit of Morozovskaia Bay [Cold Bay] are formed analogously, except that the mountains near the latter bay proceed farther northward to the coast near Isenbeck Bay [Izembek Lagoon] (only again in a N to NNE direction). And similar to the last-mentioned range, so here a lowland forms the eastern border between both of these bays. More toward the western edge of Morozovskaia Bay is the tall Morzhovskii Volcano, which reaches the approximate elevation (5,474 ft.) of the Makushin Volcano on Unalaska (this according to Veniaminov, I, p. 4; and it also applies to the Pavlovian Sopka [Pavlof Volcano]).

Between Morozovskaia Bay and Pavlof Bay the land assumes the same character again. On the west coast of the latter bay is a plateau, on which rises Pavlovskii Volcano.

At the base of Pavlof Bay there is again a lowland, 1.5 versts wide; then two lakes and a river that flows into the Bering Sea. From the east side of Pavlof Bay, toward Cape Ivanof or Kupreanof, there is on the whole the same habitus. Stepovak Bay extends close to Moller Bay, which is surrounded by mountains of low elevation.

From here on, the mountains seem no longer to appear in the same grandeur, as they recede even farther from the coast of the Bering Sea. But in the country's interior, they retain their volcanic character. And according to the testimony of the natives (Veniaminov, I, p. 231) there is a place, between 158° and 159° W. Long. and 56° N. Lat., where from the middle of a mighty mountain ridge there rises a thick column of smoke.[43] Detailed information is missing about this part of the SE coast and interior of the peninsula. We only know that NE of Cape Ayukhta [Ayutka] there is a mountainous region; and we have heard tell of Mount Chiginagak and a coast rich in glaciers, which sweeps as far as Mount Alai.

Judging by the rivers, there seem to be low mountain ranges running from SSE to NNW. On the left side of the two small rivers which flow into Kialakvit Bay [Wide Bay] there extend parallel mountain ranges. And from the sources of these rivers there probably extend flatlands of low elevation all the way to Lake Ugashek [Ugashik?], the eastern shore of which is surrounded by mountains of low elevation. The SW shore of Lake Tugat [Becharof] must be composed in the same way. And the same can be said about the area between that lake and Lake Ugashek where, according to Vasil'ev's map,[44] a river forms the connection between both lakes.

It is possible that from the NE shore of Lake Tugat, more along the interior of the land, a long range extends, the starting point of which is Four-Peak Mountain. But here, too, we find evidence of communication paths between Katmai and Svitchak [Swikshak] Bays, and Lake Iliuk. These suggest the presence of transecting valleys or high plateaus.

Finally the mountains become more significant in size on the northern side of Kamishak Bay, between Lake Iliamna and Cook Inlet, where they are pushed hard against the coast. They reach their highest elevation with the Iliamna Volcano, and they continue onward to the estuary of the Susitna River. Progressing northeast farther inland, the Chigmit Mountains run in a parallel fashion.

Although a continuous mountain chain cannot be traced along the southwestern part of the peninsula, all the known volcanos of Alaska arise on one axis line from SW to E. Long valleys, sweeping parallel to the last-named direction, cannot be found. Neither are there any rivers running in that direction. On the contrary, the direction of several mountains, (transverse) valleys, and ravines, runs SSE to NNW. East of Lake Tugat, however, the direction from SW to NE seems to be the predominant one. This would suggest at once the possibility of elevations running in two different directions: one, the direction of the North American coast range as far as Mt. St. Elias, the other one approximately at right angles to that direction, i.e., over the length of the Alaska Peninsula. This hypothesis would unite the different opinions concerning the distribution of mountain ranges.

With regard to the geologic condition of the peninsula, we refer to the conclusion of this treatise and to the attached map [see Map 5, pages 152–153]. Because of the dearth of existing material, the map can, of course, not deliver a complete geologic picture.

CHAPTER THREE

→‑ ‑←

West Coast of North America Between 59° and 69° N. Lat.

BEFORE WE LOOK AT THE ISLANDS surrounding the Alaska Peninsula, and the row of the Aleutian Islands, we want to compile whatever little material exists on the west coast of the mainland as it extends northward. This is gleaned from the geographic data of the travels of Cook, Kotzebue, Beechey,[45] Lütke, Simpson and Zagoskin, together with the material in the collections of Voznesenskii, Kupreianov,[46] Fischer and Isenbeck[47] at the Academy of Sciences.

Father Veniaminov (I, p. 106) says in the course of his demonstration, given at the end of our work, that the Aleutian Islands are not of volcanic origin, and that at one time the mainland of America and that of Asia were connected. "Elevations can be observed along the entire American coast of the Bering Sea; and they extend even to the base of the mountains located in the country's interior. One only needs to search out a slightly elevated point in order to observe from there that a sea and its waves apparently were arrested here; and that its waves were transformed into sand and mud, which is now covered with vegetation. In the midst of these petrified waves, rocky islands, some with high elevations, can be observed, which have, so to speak, been lifted up from the soil."

As we shall see in the sequel, this statement of Veniaminov's can be explained on the basis of people's myths, observations on the part of promyshlenniks and travelers, and by the occurrence of volcanic formations. As we look at the three parts of the mainland, peninsulas that reach farthest into the sea with their promontories, Romanzof, Prince of Wales, Golovnin [Point Hope] and Lisburne, we turn our attention first of all to the one between Bristol Bay and Norton Sound. The larger part of the coast of this peninsula is supposed to consist mainly of sand and mud. The northern coast of Bristol Sound rises up without any outcroppings of rock, high enough so as not to be flooded by the rising tide. But here, too, Dr. Isenbeck collected deposits of black and red lava, as well as porous basalt, which contains olivine [silica of magnesium], and andesite.

At Cape Newenham (58°42′ N. Lat.) there are two mountains (according to Cook, III, Vol. I, p. 130), one behind the other. The western, inner one of the two reaches a significant height. Farther on, Cape Vancouver is also high and rocky. The island Nunivak, which lies opposite this coast, has (according to Lütke, *p.n.*, p. 254) a continuous coast of low elevation when observed from the west. It ends with reddish-colored outcroppings. The western promontory of the island is high and steep; and in the middle of the same, gently sloping mountains of medium elevation can be observed. Farther westward from here St. Matthew Island or Cook's Gore Island (Cook, III, Vol. I, p. 164) rises from the Bering Sea. It was discovered by Capt. Siniavskii in August of 1766. According to Lütke (*p.n.*, pp. 341–343), the coasts of this island are in part high and rocky, and in part low. Eastward it ends in Cape Upright, formed like a peninsula with rocks that rise vertically to 1,400 ft. In its proximity (4 miles distant) is the location of Pinnacle Island. It is called by that name because its high summit terminates in several rocks which have the aspect of pinnacles. Twelve miles west of Cape Upright, Sugarloaf Cape rises to an elevation of 1,350 ft., in the shape of an irregularly pointed cone; the west cape, Gore, is also high. The small Morzhovoi [Hall] Island rises from the sea at the northern cape. The predominant rock is mica shist (Cape Upright). But there are many volcanic types of rock as well, and traces of volcanic activity. Before the discovery of St. Matthew Island, it was not inhabited, the same as the Pribilof Islands. Nor was it known to the neighboring peoples. This could justify the assumption that these islands emerged fairly recently.

Between the high and steep Cape Rumiantsov [Romazof] and the southern coast of Norton Sound we lack any detailed information. St. Lawrence Island, located in this latitude, "is of medium elevation," according to Chamisso (Kotzebue's first voyage, Part III, p. 170). "And the mountain ridges are flat on top. On the SW coast of the island (63°13′ N. Lat.) there is an outcropping of a greenstone-like conglomerate of a mountain. And near the south cape (62°47′ N. Lat.), as well as at the base of the same, there are large boulders of granite. The shape of the appearance of this promontory, when viewed from the sea, had aroused our curiosity. We believed that we had discerned basalt-like columns, which stood almost vertically, but all leaning toward the south." According to Beechey (German edition, I, pp. 379–383), the next mountains rise on the western point of St. Lawrence Island about 500 ft. above sea level; and they are covered with rock boulders that look like ruins when observed from on board ship. Toward S and SW these mountains terminate in steep rock walls at the shore. And offshore there are three small islands. To the north and west the mountains descend gradually all the way to the seashore. But at the NW end of the island they form a wedge-shaped promontory. The upper parts of the island were buried under snow. But the lower portions were barren (the same as on Bering Island) or covered with moss or grass. The middle of the island, when seen from the north, is so low that it appears divided into two parts; this, however, is not the case. The mountains on the eastern part of the island, which Capt. Cook named after Capt. Clarke, his associate, make up the highest part of the island. On the sixteenth of

July 1826, they were covered with deep snow. Voznesenskii collected lava specimens and basalt on St. Lawrence Island, at the west cape and in the middle of the north coast.

From Zagoskin and Voznesenskii we received some geographic data on the region, which is bordered by the south coast of Norton Sound—the Unalakleet River and the Kvikhpak (or Yukkhana, or Yuna [Yukon]) River.

The entire south coast consists predominantly of different kinds of basalt and lava. West of Cape Aziachagiak [Romanof Point] the shore is 20 ft. high between the rivers Pashtolik and Pashtoliak. There are outcroppings of grey basalt-tuff, covered with delta soil. On the 300-foot-high Romanof Point, however, sandstone can be observed (Zagoskin, II, p. 115). Farther northward, off the coast, lie the islands Steffens or St. Michael (with the fort of the same name at 63°28'45" N. Lat. and 161°44'01" W. Long.).

According to a popular legend (Zagoskin, I, p. 32), St. Michael Island was lifted up by subterranean forces. Old people claim they can remember being completely inundated by water on two occasions. At this time there rises from the island's middle a hill 300 ft. in elevation. According to Zagoskin, the island consists predominantly of basalt and lava of different kinds, which are covered with a water-rich tundra. On the north side of Cape Steffens [Stephens], north of Fort St. Michael, Voznesenskii collected black pumice stone containing lava and a reddish, hard lava with many empty bubble cavities. The northernmost promontory of the island consists of olivine-containing lava (basalt). In it there are pockets which are filled with a clay that contains a blue, basalt-like hornblende (amphibole). On the north shore of St. Michael, the so-called Shaman's Rock is composed of red, slake-like lava entirely.

Little Shelekof Island near the redoubt also consists of black lava, according to Voznesenskii, which changes to basalt-like, olivine-containing lava, and then into basalt. The olivine content is finely interspersed in large amounts throughout the rock. The coast opposite Fort St. Michael is steep, all the way to Cape Kikhtaguk, at 63°29'03" N. Lat., and 161°11'19" W. Long. (Zagoskin, I, p. 21). But it is only 20 ft. high, and contains porous basalt and lava in boulders, 10 cu. ft. in size, which have collapsed on top of each other and present the aspect of peculiar configurations. Hills 150 ft. to 200 ft. in elevation sweep parallel to the shoreline, and about 30 miles inland the Uengikhliuat [Ungalik?] Range rises. Its volcanos reach an elevation of 500 to 1,000 ft.; and their summits have kettle-shaped indentations, some of which form lakes filled with water. After Cape Kikhtaguk there follows NE a shallow bay, behind which the shoreline rises to between 40 and 60 ft. in elevation. Then at Cape Nugwulinuk [Nigvil'nuk, today Black Point] it reaches 80 ft. And here it consists of granite-like rock. In the clay-like, washed-up land within the bay of the river by this same name, ribs, shin bones, and tusks of mastadons were collected and sent to the Academy of Sciences by Mr. Voznesenskii in 1843. They seem to resemble the presently living elephant more than the mammoth.

From Cape Nugwulinuk onward the shoreline becomes lower. But then it soon rises again until it reaches an elevation of 150 ft. at Cape Tolstoi [Tolstoi Point]. From the latter there extend 200- to 300-foot-high rock walls along the coast. They consist of clay-slate; and 4 miles from the mouth of the Unalakleet (Unalakhlik) River they turn eastward inland. According to Voznesenskii there can be found in this clay-slate veins of yellow calcspar, which frequently occurs as detritus as well, with quartz, jasper, and lignite. On the banks of the Unalakleet River Voznesenskii found bituminous coal in larger amounts.

Between the Unalakleet River and the Kvitchpak (Yukkhana [Yukon]) River, there sweep the Züzeka [Zagoskin's Tstsytseka?] Mountains, which should be understood as the continuation of the last-named mountain range. It consists of clay-slate and graywacke-sandstone. This is found above Khogoltlinde Village (Zagoskin, I, p. 166). The same formations are found downstream as far as the Nulate [Nulato?] River. From there to the mouth of the Yunaka [Yukon] River (where potter's clay is found in the Unuliatcha Rock), the banks of the Kvikhpak River are covered with flood-sediment (Zagoskin, I, p. 154), wherein can be observed deposits of clay-slate, talcose slate, and marekanite.[48] Occasionally protruding rock walls consist of firm clay-rock formations, where the layering can be clearly observed. On the left side of the Kvikhpak River, 20 miles distant from its banks, there sweeps a 600-foot-high mountain ridge, which begins at the Notagash Volcano. The course of the Kvikhpak [Yukon] River follows these mountains.

At the Molekoshitno [Melozitna?] River, a left tributary of the Kvikhpak River, there is an outcropping of porphyry.

Southwest of Khogoltlinde there arises on the right bank of the Kvikhpak River, opposite the confluence of the Khutulkakat [Khotol?] River, the pointed Ilaekhogozokh Volcano. This same riverbank consists of sandstone all the way to Makaslag or Washitshagat [Zagoskin's Vazhichagat]. There follow clay stone formations from there to the Anvig [Anvik] River. Bloodstone is found there and ocher of different colorations. From the Anvig River to the mouth of the Innoko, especially between the latter and the Kvikhpak River, the terrain is low and cut through with river arms. But then, on the right bank of the Kvikhpak River up to the Ikogmiut Settlement, there rise the Ilivit Mountains (Zagoskin, II, p. 107 and I, p. 17), which reach an elevation of 2,500 ft. Basalt and lava rock are exposed at the riverbank, the same as is found near Fort St. Michael.

But 5 miles before Ikogmiut, the Chiniklik Mountains (the highest summit of these mountains) send ridges to the same riverbank. They consist of green jasper and firm clay stone, tinted red by ferrous hydroxide. According to the report of the natives, there are supposed to be cleavages of copper in the Ilivit Mountains, too. It is easy to observe how this mountain range gives direction to the flow of the Kvikhpak River. With the estuaries of the Kizhunok [Kashunuk?] and the Kipnaiak [Black] Rivers, the Kvikhpak then embraces Ingiguk [Kusilvak] or Maggemiutskoi Mountain, and then Ingychuak Mountain [Ingrichuak Hill], the continuation of which sweeps along the right bank of the Kvikhpak River northward. It is supposed to consist of basalt and clay-slate. Along the Ingiguk River the Agulymiut people collect iron phosphate of dark-blue coloration,[49] slaty texture, and firm consistency. Potter's clay is obtained from the region between the Kizhunok and Kuskokwim [Rivers]; also cimolite, and red and black bole.

We also observe that the margins of the entire, perhaps crater-shaped region consists predominantly of volcanic rock and shale formations. Granite-like rock, however, occurs only in one place, and that rather doubtful. Conditions are different at the banks of the Kuskokwim River.

East of Fort Kalmakof [Kolmakof], on the right bank of the last-named river, there rise the Tashatuling or Tashatulig [Kilbuck?] Mountains, which reach an elevation of 2,000 ft. They consist predominantly of mica-rich granite (Zagoskin, II, p. 14). And downriver, too, the right, mountainous bank of the Kuskokwim, opposite the mouth of the Tulukagnac River, is composed of granite, judging by the boulders that are scattered about. But the left bank is flat. Only 20 miles inland is a mountain ridge that reaches 2,000 ft. in elevation and forms the watershed between the Kuskokwim and Nushagak Rivers.

On the lower run of the Kuskokwim River, people gather pumice stone, red ocher and cimolite. And on the upper run there are four mountain chains, sweeping from E to W, which are separated by the Tochotno [Takotna?] River (a tributary from the right side of the Kuskokwim River). From their needle-shaped, rugged mountain peaks, Zagoskin (II, p. 101) deduces their volcanic origin.

Our knowledge about the geographic conditions of the inner coast of Norton Sound is very incomplete, to say the least (cf. Lütke, *p.n.*, p. 249–252). Cape Denbigh forms a peninsula where, according to Zagoskin (II, p. 118), there are outcroppings of porphyry. It is separated from the mainland by a low, sandy isthmus. Golovnin Bay is bordered by two high capes, Cape Darby and Cape Kamennoi (Stoney Cape) [Rocky Point]. The latter is the lower of the two; it got its name because it consists of a bare, flattened rock. A river terminates with five arms into the back region of the bay. It seems to be connected to Lake Imuruk of Grantley Bay [Grantley Harbor]. Then there follows a mostly flat coast until the low Cape Rodney, behind which high, snow-covered mountains rise upward, 2 miles inland from shore. Only the two promontories along this stretch—the Utesistye (rocky) Foothills [Topkok Head?] and Tolstoi Foothills [Cape Nome?]—consist of steep rock walls. Off the coast, between the capes Tolstoi and Rodney, lies Asiiak Island (Cook's Sledge Island, cf. his *Third Voyage*, I, p. 140). Its circumference is 12 miles. Its elevation is low, and it is covered with large rock boulders. Teben'kov, however, (Lütke, *p.n.*, p. 247) describes it as a rock column, 1 mile in circumference, which protrudes 642 ft. above sea level. It is possible that Asiiak was confused with Ukivok in that report.

The latter island, or Cook's King's Island, is located opposite the low coast, where the land rises in the background, between Cape Rodney and Port Clarence. According to Kotzebue (first voyage, Vol. I, p. 138) it is a rock of 585 ft. in elevation; and according to Beechey (cf. also Lütke, *p.n.*, p. 248) its elevation is 756 ft., and its circumference 1 mile.

Port Clarence consists of two wide basins (cf. *Zoology of Capt. Beechey's Voyage*, 1839, Appendix: Geology, p. 179). The outer one's southern limit is a long spit of land, Point Spencer, which is covered with alluvial deposits. Between there and the northern side of the bay, the entrance is only 3 miles wide. The inner basin is called Grantley Harbor. This is bordered by similar spits of land on both sides. The coast of the former basin is mostly of low elevation and features many lakes and lagoons. But on the northern and eastern side the coast changes its character; and the north, south and east shore of Grantley Harbor is mostly steep and rocky. The rocky shore at Cape Riley extends for 2 miles; and then it blends on both sides into the low-lying coast. It consists of loose mica and talcose slate with veins of calcite that glistens like mother-of-pearl, as well as grey quartz. It seems that this slate falls in from the NE. At the coast of the inner harbor (Grantley), at about the middle of the southern side, there is an outcropping of similar mica schist and alum schist, with an eastern gradient. And the same combination and direction of layering has been observed at the north coast where black-colored crystals and masses of chlorite break in. The head of the harbor also seems to feature the same formations.

From Port Clarence on to Cape Prince of Wales, or Nykhta, a mountain range is formed along the coast [York Mountains], which reaches an elevation of 1,876 ft. at a location 3 miles from Jackson Peak at 65°30′ N. Lat. and 167° W. Long. Two mountain ridges extend from this range in an ENE direction. The first of these features the 2,597-foot Snow Mountain (Beechey, II, p. 350). The second one ends at Cape Wales. The coast in front of these promontories is bordered by steep, rocky cliffs, and by deep ravines. Cape Prince of Wales itself is a rock column, according to Chamisso (in Kotzebue, III, p. 170). According to Beechey (l.c.) it is a mountain, covered with rock boulders, from which extends a wall-shaped ridge of naked rock, which appears in peculiar, irregularly interrupted shapes. This wall begins at the NE base of the mountain, ascends to its peak, and re-emerges on the mountain's southern and southwestern side, albeit more separated and less conspicuous.

East Cape, too, which lies diagonally across from there on the Asian coast, consists of an abruptly descending, rocky peninsula. In front of its point there are a few rocks that resemble church steeples (Cook in the German edition, I, p. 147). We will quote Kotzebue's more involved description of this promontory (I, p. 156). "On the outermost point of Cape Oriental there is planted a sugarloaf-shaped mountain on the low land. It rises vertically upward from the sea, and its peak is caved in,[50] and is open toward the seaside. This location has a dreadful appearance because of the black rocks that have collapsed in a heap of confusion. One of these, shaped like a pyramid, is particularly conspicuous. These dreadfully veined rocks make one mindful of the revolution of the earth, which had happened here at one time. But for that, Asia and America were once connected. The appearance, as well as the position of the coast makes this plausible. And the Gvozdev or St. Diomede Islands are remnants of the connection between the East Cape and Prince de Galles." The southernmost of these islands, called St. Diomede by Bering, according to Beechey (German ed., I, p. 386), is Fairway Rock or Ukiyak (Okivaki, according to Sauer, p. 258). It is a tall, four-cornered rock. The island in the middle, Krusenstern Island or Inaklit [Little Diomede], has vertical rock walls and a flat surface on top. The third, largest, and northernmost of them, Ratmanov or Imaklit Island (Imaglin according to Sauer) [Big Diomede], is 3 miles long and 10 to 15 miles in circumference. Toward the south it is high; and on the opposite side it ends in low shore-walls, with small, pointed cliffs in front of them. More accurate information about these islands and the corresponding coastlines we have not been able to find. But most of the argonauts agree by and large that the shores of Bering Strait at this latitude resemble each other a great deal. We will, however, return to the subject.

After Cape Prince of Wales there follows a low-lying coast. Behind it is a sandy ridge of land. At Shishmaref Bay, Voznesenskii collected lime marl with fragments of a bivalve (*Modiola?*), the epidermis of which is completely preserved. And on the coast of the Arctic Ocean between Shishmaref Inlet and Cape Espenberg, in the region of Devil Mountain (on the nineteenth of July 1843), he found coarse yellow limestone with veins of white calcite, lime marl, black clay-slate with a content of lime, mica schist, and a breccia of feldspar, hornblende and mica with a lime-like agglutinant.

The peak of Devil Mountain, 616 ft. high (trigonometric measurement according to Beechey, II, p. 450), has a considerable circumference, according to Kotzebue (Part I, p. 148). People perceive it as the ruins of a destroyed castle, with only a few towers left standing. These, however, are volcanic rock formations. This is apparent because of the lava stream, which extends from Devil Mountain to the sea (according to the geological portion of Beechey's *Voyage*). Kotzebue Sound beginning with Cape Espenberg, was geognostically analyzed and graphically depicted by Buckland, Belcher, and Collie (*The Zoology of Captain Beechey's Voyage,* Appendix, p. 169, Pl. I). We take up that work in this place and add a few new data.

Kotzebue Sound consists of three bays: Goodhope Bay, Spafarief Bay, and Eschscholtz Bay. They follow each other from west to east. In front of the latter bay there lies Choris Peninsula and Chamisso Island.

On the west side of the east cape of Goodhope Bay, called Cape Deceit, there is a high rock cliff. Here the coast consists of dark blue slate and lime shale, with seams of mica schist layered in between. The former bears strong traces of the influence of fire. Gullhead [Toawlevic Point], a narrow, rocky peninsula that extends 1 mile into the sea, consists mostly of calcareous slate of a blackish or greyish color, which contains selenite ("St. Mary's glass"). Rock formations of medium elevation alternate along this southern coast of the bay with gentle slopes for a distance of 8 miles northwesterly, with the Devil's Ears in the background. But then the coast takes on a different aspect, as it features many small promontories and coves, for the distance of about four miles. That coast lies low, and rises gradually inland, where vegetation is apparent. The prominent points of this coast are completely covered with large boulders of porous, vesicular, and compact lava, wherein refracts olivine. A few of the lava boulders are partially immersed in water; others are surrounded with masses of sand, and again others stand there free and isolated. On some of the boulders, hollows the size of a fist can be observed; and a black volcanic sand covers the coast all the way to the narrow Cape Espenberg, which consists of tall sand dunes. From that point onward no more boulders can be found. In the proximity of the cape just mentioned, Voznesenskii collected *anthrakonit* [anthrachinon?] and black, crystalline marble. Beechey (II, p. 41) found in the dark-colored, volcanic sand the shells of *Cardium, Venus, Turbo, Murex, Solen,* and *Tellina,* as well as a few large *Asteria.*

Cape Deceit (l.c., p. 37) seemed to consist of large, jagged boulders of coarse limestone, which presented no clear layering. East of there, at the second promontory, and farther on between the estuaries of two unnamed rivers, there are deposits of calcareous slate with contents of selenite (lady's glass). And on the most weathered parts of the rock bank there was talcose slate with thin layers of limestone in between, while the steeper cliffs were composed of coarser limestone. At one location dark-blue clay-slate was also observed. At the mouth of the river at the farthest recess of Spafarief Bay there are mud-cliffs alternating with alluvia. They continue (farther eastward) along the northward-rising coast of the bay toward a 640-foot-high mountain (Beechey, I, p. 468). From there on to Eschscholtz Bluff, and a little beyond, the shoreline consists of mica schist, clay-slate, and foliated chlorite with inclusions of quartz, calcspar, chlorite, feldspar, tourmaline, garnet and pyrite. Within these shales, which change and blend with one another, there occur layers of blue and white primitive limestone with slaty texture and strong westward slope. But at Eschscholtz Bluff there predominates chlorite-slate with pyrite. Eschscholtz believed he found widely extending ice or glacier masses between the last-named bluff and Elephant Point (Kotzebue, I, p. 146 with illustration, and III, p. 170; Gilbert's *Annales* IX, 1821, pp. 143–146). But according to Beechey, these are merely surface coverings or frozen earth. This subject is of enough interest to warrant addressing Beechey's, Belcher's and Collie's comments in this place (Beechey, German ed., I, pp. 403–406, and II, pp. 32–34; Eng. ed. pp. 257, 323, 560; and Appendix by Buckland, pp. 593–613 [Eng. ed. London, 1831, vol. I, pp. 352–354, 443–445]).

While the duties of the ship were being forwarded under by first lieutenant, Mr. Peard, I took the opportunity to visit the extraordinary ice-formation in Escholtz Bay, mentioned by Kotzebue as being "covered with a soil half a foot thick, producing the most luxuriant grass," and containing an abundance of mammoth bones. We sailed up the bay, which was extremely shallow, and landed at a deserted village on a low sandy point, where Kotzebue bivouacked when he visited the place, and to which I afterwards gave the name of Elephant Point, from the bones of that animal being found near it.

The cliffs in which this singular formation was discovered begin near this point, and extend westward in a nearly straight line to a rocky cliff of primitive formation at the entrance of the bay, whence the coast takes an abrupt turn to the southward. The cliffs are from twenty to eighty feet in height;[51] and rise inland to a rounded range of hills between four and five hundred feet above the sea. In some places they present a perpendicular front to the northward, in others a slightly inclined surface; and are occasionally intersected by valleys and water-courses generally overgrown with low bushes. Opposite each of these valleys, there is a projecting flat piece of ground, consisting of the materials that have been washed down the ravine, where the only good landing for boats is afforded. The soil of the cliffs is a bluish-coloured mud, for the most part covered with moss and long grass, full of deep furrows, generally filled with water or frozen snow. Mud in a frozen state forms the surface of the cliff in some parts; in others the rock appears, with the mud above it, or sometimes with a bank half way up it, as if the superstratum had gradually slid down and accumulated against the cliff. By the large rents near the edges of the mud cliffs, they appear to be breaking away, and contributing daily to diminish the depth of water in the bay.

Such is the general conformation of this line of coast. That particular formation, which, when it was first discovered by Captain Kotzebue, excited so much curiosity, and bore so near a resemblance to an iceberg as to deceive himself and his officers, when they approached the spot to examine it, remains to be described. As we rowed along the shore, the shining surface of small portions of the cliffs attracted our attention, and directed us where to search for this curious phenomenon, which we should otherwise have had diﬁculty in finding, notwithstanding its locality had been particularly described; for so large a portion of the ice cliff has thawed since it was visited by Captain Kotzebue and his naturalist, that only a few insignificant patches of the frozen surface now remain. The largest of these, situated about a mile to the westward of Elephant Point, was particularly examined by Mr. Collie, who, on cutting through the ice in a horizontal direction, found that it formed only a casing to the cliff, which was composed of mud and gravel in a frozen state. On removing the earth above, it was also evident, by a decided line of separation between the ice and the cliff, that the Russians had been deceived by appearances. By cutting into the upper surface of the cliff three feet from the edge, frozen earth, similar to that which formed the face of the cliff, was found at eleven inches' depth; and four yards further back the same substance occurred at twenty-two inches' depth.

The glacial facing we afterwards noticed in several parts of the sound; and it appears to me to be occasioned either by the snow being banked up against the cliff, or collected in its hollows in the winter, and converted into ice in the summer by partial thawings and freezings—or by the constant flow of water during the summer over the edges of the cliffs, on which the sun's rays operate less forcibly than on other parts, in consequence of their aspect. The streams thus become converted into ice, either while trickling down the still frozen surface of the cliffs, or after they reach the earth at their base, in which case the ice rises like a stalagmite, and in time reaches the surface. But before this is completed, the upper soil, loosened by the thaw, is itself projected over the cliff, and falls in a heap below, whence it is ultimately carried away by the tide. We visited this spot a month later in the season, and found a considerable alteration in its appearance, manifesting more clearly than before the deception under which Kotzebue laboured.

The cliff we had ascended was composed of a bluish mud and clay, and was full of deep chasms lying in a direction parallel with the front of the eminence. In appearance this hill was similar to that at Elephant Point, which was said to contain fossils; but there were none seen here, though the earth, in parts, had a disagreeable smell, similar to that which was supposed to proceed from the decayed animal substances in the cliff near Elephant Point.

I found Mr. Collie had been successful in his search among the cliffs at Elephant Point, and had discovered several bones and grinders of elephants and other animals in a fossil state....

The cliff in which these fossils appear to have been imbedded is part of the range in which the ice formation was seen in July. During our absence (a space of five weeks) we found that the edge of the cliff in one place had broken away four feet, and in another two feet and a half, and a further portion of it was on the eve of being precipitated upon the beach. In some places where the icy shields had adhered to the cliff nothing now remained, and frozen earth formed the front of the cliff. By cutting through those parts of the ice which were still attached, the mud in a frozen state presented itself as before, and confirmed our previous opinion of the nature of the cliff. Without putting it to this test, appearances might well have led to the conclusion come to by Kotzebue and M. Escholtz; more especially if it happened to be visited early in the summer, and in a season less favourable than that in which we viewed it. The earth, which is fast falling away from the cliffs—not in this place only, but in all parts of the bay—is carried away by the tide; and throughout the summer there must be a tendency to diminish the depth of the water, which at no very distant period will probably leave it navigable only by boats. It is now so shallow off the ice cliffs, that a bank dries at two miles' distance from the shore; and it is only at the shingly points which occur opposite the ravines that a convenient landing can be effected with small boats.

Three possible explanations emerge from Beechey's and Collie's reports concerning the ice formation above these steep shorewalls (Beechey, *Voyage*, Part II, Appendix): they could develop 1) as a result of snow heaping up and gradually turning to ice in the ravines along the shoreline; 2) through water freezing in the rifts and hollows (which are created by the collapsing of the frozen mud); 3) by water seeping into the top part of the shorewall, and freezing in the process.

It is most likely that all three processes were at work in forming those masses of ice. Whatever the processes which caused formation of the ice, the fact remains that Kotzebue and Eschscholtz were mistaken when they thought the shorewalls of frozen mud, clay and sand were true icebergs, although they saw much more ice there than did Beechey.

Concerning the fossil remains at Elephant Point, we have some more information in Buckland's edition of Collie's data (Beechey's *Voyage*, Part II, London, 1831, 4°, Appendix pp. 593–613, entitled: "On the occurrence of the remains of the elephants and other quadrupeds in the cliffs of frozen mud in Eschscholtz Bay, within Bering Strait, and in other distant parts of the shores of the Arctic Sea," with III Plates). The 2.5-mile-long, 90-foot-high coast consists of deposits of clay and very fine quartz and mica sand, which is a grey color when it is dry. At the base of these deposits, the bones are washed up by the waves of the sea.

And the ebb tide takes them out to sea. Thus a sandbar is formed 50 to 100 yards off the coast, from where the collected pieces were taken. Noteworthy, furthermore, is that these fossil remains are not embedded in mere ice. They are instead covered with mud and sand. And they do not occur on top of the steep cliff, nor generally in the upper layers. An unpleasant odor of burnt bones spreads over the site of the find of these bones. The same odor, however, was also noticed at a location east of Elephant Point, where no such remains were found. Such remains were found at any rate only in one more location within Kotzebue Sound, i.e., at a 50-foot-high cliff at Goodhope Bay.

Buckland (Pl. I and II) defined the mandible and the large and small tusks of an extinct species of elephant (after Cuvier's *Ossemens fossiles*, Vol. I, Pl. II, Fig. 1, 4, 5; Pl. V, Fig. 4, 5; Pl. VIII, Fig. 1; Pl. IX, Fig. 8, 10; Pl. XI, Fig. 2). They are similar to the skeletons provided by Adam in the Academy of Sciences at St. Petersburg. Buckland adds the remark that the strongly curved, twisted teeth are often found on fossilized elephants, but not as consistently as with live elephants of Ceylon. On Plate II, there are furthermore depicted the thighbone (femur), the epiphysis from the lower part of the same, the shinbone (tibia), the shoulder blade (scapula), the os innominatum, and the os calcis of elephants. Plate III shows the head of *Bos urus* (the head and other parts of the bison are not depicted). There are fragments of antlers of a species of deer, reindeer perhaps. And there are the tibia and radius of a large kind of deer; also astragalus, metacarpus, and metatarsus of a horse, and a vertebra of an unknown animal. Buckland proves that such fossil remains are found all over Europe and northern Asia in the same kind of diluvial soil. Only in this arctic region, the temperature was particularly advantageous for their preservation. Later on we shall return to Buckland's further remarks, which are based on these discoveries.

At the base of Eschscholtz Bay, the hills rise no higher than 600 ft. to 1,000 ft. in elevation. The coast at the mouth of the Buckland River consists of alluvial deposits and mud-cliffs. The same is true of the northern shore of Eschscholtz Bay; only there, the elevation is no higher than 40 ft. The Choris [Baldwin?] Peninsula, which follows that bay, is divided into two parts by a lowland. On the northern, more rounded and cone-shaped part, a hat-shaped peak (Hat Peak) rises 600 ft. above sea level. On the west side of the southern portion of this peninsula, the rocky shore consists of uninterrupted deposits of mica schist with veins of quartz and feldspar, 150 ft. to 200 ft. high, which fall in under 30° W, refracting also schorls, garnets, hornblende, and calcspar. The east side is not as high or broken up. And the grey mica schist, fragments of which lie about everywhere, does not permit a clear observation of any layering. At the first, the southernmost promontory, directed toward Eschscholtz Bay, the deposit layers fall in from 60° NE. And in the mica schist here, there are garnets, veins of feldspar with schorl, and layers of quartz. At about the midpoint between this promontory and the lowland, a seam of milk quartz comes to the surface. This location is conspicuous because of the white boulders that have broken off and fallen here. A little farther northward, a deposit of limestone, 10 yards high and 5 yards wide, covers the mica schist. It extends all the way to the west side of the peninsula. In that place it forms the first layers of rock at the southern side of the lowland; also four steep promontories, which are divided by narrow coves. White and blue layerings can be observed here, falling in with no more than 5°. In the upper strata of this limestone, deposits of pyrite are found, and cavities filled with chlorite-soil. In the lower layers there are often deposits of mica with hornblende, either coarse, or crystallized in flat prisms, also tourmaline, garnets and pyrite. At times the quartz deposits take on the coloration of topaz, and often this is surrounded by only small quantities of chlorite. In one location, the pyrite forms a continuous stratum. In one of the foothills there is a deep and spacious cave.

Off the southern point of the Choris Peninsula, Chamisso Island rises 331 ft. above sea level (Kotzebue, l.c., and Beechey's *Voyage*, I, p. 399 and II, p. 350). Its circumference is three to four miles; and it is surrounded by steep rock cliffs, except for its eastern part where it ends in a low spit of land. Its surface is hilly and furnished with lakes and brooks. In the middle of the island, a wall of bare rock forms the highest point. South of there it appears as if human hands have laid down a rock pavement in the form of an arc (shore ledge). The predominant rock is mica schist, which gradually becomes quartz-shale, and on the N and SW sides gneiss. The strata arise under 60° N, and in them refract garnets, schorl with chlorite. Hornblende, quartz, hornstone, and feldspar are also found in some strata. According to Chamisso (Kotzebue, Part III, p. 170) there is supposed to be a steep rock, Puffin Island, standing apart from Chamisso Island. It is composed of weathered granite, so much of which has already fallen off that the remaining part has the appearance of a tower.

At the entrance to Hotham Inlet there rises in the background a range of mountains (with Deviston Peak) [Deviation Peak?]. Beyond this there follows another range that sweeps along the coast past Cape Krusenstern (where Dr. Fischer collected petrified clay and light, porous, red lava) to Mt. Mulgrave. Farther northward the coast is reported to consist of alluvial land up to Thompson Promontory. The northern point of the two-part Cape Thompson, which is the higher one, rises to almost 400 ft. It consists of mountain limestone, which alternates with deposits of silica, from 6 inches to 2 ft. thick, which fall in from the west at 10°. Following this, on the part of the slope that juts out first, there is blue and black clay-slate in horizontal deposits. This alternates with limestone, 6 to 8 ft. thick, until the former predominates in strata filled with petrifications [fossils]. Into these shales a river has dug its bed. But the deposits on its left bank no longer have that fibrous, twisted appearance. And on the second outcropping they are again covered with layers of limestone and flint with a slope (dip) of about 5°. Beneath these deposits there again emerge curved strata of slate at the east slope or at the base of the second point. These contain strata of calcspar, pyrite, knolls of septaria (dentalium), and coarse limestone with tubiporia [?] and encrinites.

At the end of the bay there again emerges limestone similar to that at the outcropping. But the silicas contained in it have a grey coloration; and they contain an abundance of fossils (bivalves). In horizontal parts of these deposits there is clay-slate, covered in a few locations by efflorescences which are probably the result of deterioration of pyrite. Also Ca, C, and gypsum were observed here.

The specimens are described as follows: coal or Derbyshire limestone. The same with deposits of silica, and the latter alone. Limestone with productus and encrinites. Calcareous slate with a glass-like structure and contents of gypsum crystals. Coral-lime, tubiporous lime. Black slate and coral-limestone intermingled or alternating with each other, with calcspar and gypsum crystals. Coarse limestone with veins of calcite, spathic iron ore and zinc blende; balls of clay ironstone occurring in layers of silica. The limestone (according to Buckland) is at times indistinguishable from the entrocite lime at Derbyshire. It is composed entirely of fragments of encrinites. The same is the case with the coral: *Lithothaminion* or *Madrepora basaltiformis* and *Flustra*. The *Productus martini* also resembles this mountain limestone. Mr. Collie also mentions the occurrence of trilobites in his notes; but among the specimens there were none.

North of Cape Thompson the coast protrudes in a low point (Cape Golovnin?) about 20 miles into the sea. Volcanic sand is found in this place. Noticed on the roof of huts, to weigh them down, were blocks of clinkstone. From this point onward (Cape Golovnin or Hope), a low coast, covered with alluvial land, reaches as far as a good-sized river and a rock cliff. This cliff consists of basalt, according to Mr. Elson.

Farther northward of this point, or SW of Cape Lisburne, the grey-brown and black deposits of the steep embankment fall in toward S and W, mostly in steep angles (*Zoology of Capt. Beechey's Voyage*, p. 172). Toward the interior, the entire region rises up several hundred feet in saddle-shaped hills with wide valleys, cone-shaped elevations and steep rock walls. The latter seem to consist of mountain limestone, and the slopes of shale and clay.

Concerning Cape Lisburne itself, we hear from Beechey (Vol. I, p. 421, Eng. ed., London, 1831, vol. I, pp. 368–369) as follows:

We landed here, and ascended the mountain to obtain a fair view of the coast. . . .
Our height was 850 feet above the sea, and at so short a distance from it on one
side, that it was fearful to look down upon the beach below. We ascended by a

valley which collected the tributary streams of the mountain, and poured them in a cascade upon the beach. The basis of the mountain was flint of the purest kind, and limestone, abounding in fossil shells, enchinites, and marine animals.

According to the appendix to *The Zoology of Capt. Beechey's Voyage*, p. 172, Cape Lisburne consists of two outcroppings. The southwestern one rises abruptly; it is covered with grey rock boulders and without the slightest trace of vegetation. The higher NE cape rises gradually to an elevation of 850 ft. above sea level. With its green covering, however sparse, it offers a striking contrast to the grey peak of the other one. The first point rises from the sea with clearly discernible layers, which fall in at a southwesterly angle (58°). They consist of bituminous limestone in the middle, but along the sides of slightly deteriorating slate and clay. On the rough face of the second point, the stratification was not traceable. In part it is covered with vegetation or with fallen-down heaps of flint that also contains bituminous limestone. Mountain ranges extend southeastward from both points.

The specimens (l.c.) were defined as follows: black clay-slate with fragments of *Terebratula* in thin deposits from the NE side of the first cape; *Tubipora* in black bituminous stone from the same location; terebratulae and a radiated encrinite head in rich, dark bituminous stone; column-shaped madrepores.

From a black clay-slate with lime content (cf. Appendix I), Dr. Fischer and Kupreianov collected *Cyathophyllum flexuosum, C. caespitosum, Turbinola mitrata, Cariophyllia truncata, Sarcinula* (sp.?), Cyathocrinites (?), Brachiopoda (Gen.?) in coarse grey limestone; veins of calcite in clay-slate, itacolumite (flexible sandstone), quartz rock, jasper, clay-ironstone, flint, conglomerate of siliceous limestone (boulders).

Northeast of Cape Lisburne no mountain ridges can be seen, nor any highland; and the low coast of the lagoon slopes deeply and widely for a distance of 50 miles. Forty-five miles from Cape Lisburne a ridge of hills sweeps northeasterly. It consists of sandstone, ranging south-southwesterly, 25°. Its gentle slopes are clothed in green. The ridge of the range consists of barren layers of rock. On the NW side they rise up vertically for several feet. Beneath similar vertical layers of the NE slope, coal lies exposed, which is mixed with components of soil.

Transected in several places by wide valleys, where there are lakes and rivers (Beechey's *Voyage*, I, p. 422), this elevation rises up near Cape Beaufort to become a high mountain range (*Zoology of Capt. Beechey's Voy.*, l.c.). In it, exposed one-quarter mile from shore, in an ENE to WSW direction, is a narrow seam of bituminous coal. This coal is slaty; and where it crops out, it is dry and poor. But a few small pieces, dug out by a small animal, most likely an ermine, burned quickly with a bright flame.

The upper part of Cape Beaufort rises steeply from the sea and is covered with debris of a shaly sandstone. Contained therein are grasses both ripened and smooth, petrified into the coal. Between the thin layers of sandstone, mostly of a grey-red color, there are lumps of hornstone, quartz, clay-ironstone, siliceous or Lydian limestone, with veins of calcspar. Cape Beaufort is the highest point of this region. It rises 300 ft. above sea level and seems to form the boundary between the first-mentioned row of hills in the SW, and the lake- and lagoon-covered lowlands, which extend northeastward as far as the eye can see.

With these lowlands there begins a diluvial soil, which extends from Cape Beaufort to Icy Cape, and to the Reindeer Station and Wainwright Inlet. Beyond the latter, the coast and the country's interior are composed of similar material. And at Cape Smith (71°13′ N. Lat., 156°45′ W. Long.), ice cliffs were observed, similar to those at Cape Blossom (Beechey, II, p. 43), and in Eschscholtz Bay.

At the Reindeer Station, boulders of granite were collected, as well as syenite, aventurine, coal, and many specimens of hardened clay. And under 71° N. Lat. and 162° W. Long., the dredge brought up from the sea bottom: grey sandstone, hard clay and coals in large amounts, and hard clay with imprints of plants.

A further pursuit of the geologic and geographic conditions of the polar coast of America lies beyond the scope of our assignment. We therefore refer the reader to the geoglogical essay by Richardson in Appendix I to Franklin's *Second Expedition*. Later on we shall return to give an overall assessment of the coast described by us above, as now we shall return to a description of the most important islands in the region of Alaska.

CHAPTER FOUR

→> <←

The Most Important Islands in the Region of Alaska

KODIAK ISLAND EXTENDS from 56°45′ to almost 50° N. Lat., parallel to the coast of Alaska. It is separated from the same by Shelikof Strait. Northeastward from there is the narrow Kupreanof Strait, which divides it from Afognak Island.[52] Thereafter follows Shuyak Island; and farther NE, toward the Chugach [Kenai] Peninsula and its Cape Elisabeth, there are the Barren or Peregrebniye Islands and the Chugach Islands. Other islands in the vicinity of Kodiak are the following: Yavraschitchey [Evrashichei, today Marmot], Elovoi [Spruce], Ugak, Saltkhidak [Sitkalidak], Sitkhinok [Sitkinak], and Tugidak. Southwesterly Kodiak ends with the Trinity (Troitskii) Promontory, whereupon follow the islands of the same name. All of Kodiak is equally covered with mountains, according to Golovnin (I, p. 194). Some of these are very tall and covered with permanent snow. Between them are many broad valleys and rivers.

This island is famous for its tall trees. This fact prompted the first Russians as they advanced from the west, and Shelikhov in particular, to found here the first headquarters of the Company (cf. Sauer's depiction, p. 182, and Sarychev, II, p. 36). The climate is the same as that of the SE coast of Chugatsk [Kenai Peninsula?], i.e., milder and drier than it is on the rest of the islands. More detailed information can be found in Baer's and Helmersen's *Beiträge*, Vol. I, Art. XI.

According to Golovnin (I, p. 194), and Voznesenskii, the island is mountainous, but not very high;[53] nor are there any volcanos. Along the coast, it consists predominantly of black clay-slate, of which we have specimens from Pavlov Harbor [Saint Paul Harbor] and from the settlement nearby. Opposite the former, the little island, Dolgoi [Long], consists entirely of clay-slate, with slickensides. On the NW shore of this island the clay-slate forms pyramid-shaped rocks, while on the western side it is horizontally stratified. We received (from the Karluk settlement on Kodiak's west coast) graphite in small pieces, which probably breaks in deposits of lime, as we conclude by analogy. The people from Kodiak obtain amber from Alaska by barter. They also need sulfur, as do all the other tribes of this region (Pallas in Busching's *Mag.*, XVI, p. 276). They powder a beaver skin with it, onto which they direct sparks from their flints.

There is iron ocher, which the women use to color furs. There are furthermore many petrified wood-trunks (species of *Pinus*), especially on high ground.
They are permeated with ferric hydroxide and penetrated with pyrite. One of
them is conspicuous among the specimens sent by Voznesenskii, in that one piece
of carnelian in it has retained the structure of the wood completely. This transformation of vegetable matter into carnelian has heretofore not been observed,
so far as I know, although Gaultier de Claubry found in it a stain of organic substance (*Ann. de Ch. et de Ph. L.*, p. 438 and Poggd's *Ann.* XXVI, p. 562). Carnelian is long known as a means of fossilization of animal matter (sea shells from
Blankenburg in the Harz Mountains, and madrepores [starfish] from Linder and
Tommis Mountain near Hannover, etc.). To explain the process of the carnelian-
ization on the whole, as well as that of silicification in particular, that is, the
absorption of Si from the exterior inward, it is as follows: the medium for the
solution of Si which penetrates the wood material of Kodiak cannot be found in
hot springs (as is the case in Iceland). They are not known on this island. Therefore here, as well as especially on Unga Island (cf. below), it is possible that silicification depends upon a contact-petrification with the surrounding rock; or it
develops by way of impregnation by cold, silicon-containing water. Deposits
with contents of fossils seem to be more commonly found all over Kodiak. From
a 1.5-fathom-high shore cliff, consisting of volcanic tuff, which lies north of
Tonkii Cape [Narrow Cape] (Igatskoi Bay [Ugak Bay]), Voznesenskii collected:
*Mytilus middendorffi, Mya crassa, M. truncata, M. arenaria, Pectunculus kasche-
warowi, Cardium decoratum, C. groenlandicum* and *Crassatella*. On the opposite
north coast of the island, near Uganik Village, there is: *Cardium decoratum, C.
groenlandicum*, and *Ostrea longirostris*. The former are the same as on the southern coast of the island, as well as on Alaska, Unalaska, and Atka. All seem to
belong to the most recent Tertiary epoch (cf. Appendix I).

Opposite Igatskoi Bay there lies the island Ugak, where, according to Voznesenskii, there are many deposits of jasper-clay. Thereafter follows the island
Ukamok (Chirikof or Vancouver; cf. Sarychev, Golovnin, Lütke). Its aspect
resembles the Highland in the Bay of Finland (Golovnin, I, p. 178); and it extends
SSW to NNE, has a length of 9.5 miles and a width of 3.5 miles. Its southern end
is high, and it has low points in the north. Not far from its western end there are
rock cliffs and a rock called "Nagai," on the NE side of which a row of cliffs
extends 3 to 4 miles almost parallel to the coast. They are partly submerged and
partly above water. The village is situated on a narrow bay, surrounded by rock
walls in 55°48′ N. Lat. This island could well be counted among the Evdokeev or
Semidi Islands, which are mountainous, according to Sarychev. The largest of
them, Simedan or Semidun [Aghiyuk], consists of granite rock (according to
Sauer in Billings's exped., p. 200).

The **Shumagin Islands** (Aleut: Kagigyun) were discovered by Capt. Bering on his second voyage, and named after the sailor, Shumagin, who was buried on Nagai Island. Voronkovskii surveyed them in 1837. Father Veniaminov described them, and later also Zagoskin. The latter suspects that they were separated from the mainland by a strong earthquake. Postels, too, mentions (in Lütke's *Voyage*, Vol. III, p. 26, or Berghaus's *Geography and Ethnography*, II, p. 744) that volcanic phenomena occurred here in the old times. Sauer (l.c., p. 199, Eng. ed., London, 1802, p. 167) comments on these islands as follows: "All of them are high and barren, exhibiting a great similarity in their appearance, though of various forms and sizes. . . . Some (of the mountains) project into the sea in rugged cliffs; some are sharp capes, and often terminate in bluff heads. There appear some convenient coves; but it would be hazardous to enter them, on account of the detached and sunken rocks that are scattered about. . . ."

Unga is the largest among the fifteen larger and seven smaller islands of this group. It is the most westerly situated island. Its northern end lies under 55°37′, and the southern end under 55°11′ N. Lat. Its length is 26 miles. It is poor in lakes; but there are ten rivers. It is mountainous; and on its south side especially, the coast is steep and rugged. On the NW side there extends a flatland, which ends in the moderately elevated Cape Tonkii [Unga Spit]. On Unga as well as on the other islands of this group no rocks can be seen which terminate in peaks or needles; nor are there volcanos with craters. Veniaminov mentions (I, p. 27) an enormous deluge which afflicted Unga on the twenty-seventh of July 1788, and cost many Aleuts their lives. It seems as if this flood-like deluge went from Sanak past Unga to Alaska.

On the southward-situated Ocheredinskii Bay [Acheredin Bay] the shore consists of clay-like material, colored from ocher-yellow to blood-red (*krovavik*, bloodstone). The rock walls at the so-called Harbor Bay [Delarof Harbor?], the eastern bay, where lies the Delarovskoe Village [Unga] (Uguagak in Aleut), consist of a conglomerate of siliceous limestone with a clay-sandy binder. Predominantly distributed on the island are horizontally deposited strata of clay and sand. Among them is found a sandstone similar to itacolumite. This gave cause to the fallaciously disseminated rumor that diamonds were to be found on Unga.[54] On the northwest side of the island, in the proximity of Sakharovskaia Bay [Zachary Bay], on the west side of the same, there is a collapsed cliff where at 200 ft. elevation (50 *Tois.* = 320 ft., according to Lütke) above sea level, four layers of lignite are found between clay and sand (bituminous wood, brown coal to hard coal, 1 to 2 ft. in thickness, according to Voznesenskii). It was found to be useless for the manufacture of cast iron, according to the tests conducted in New Archangel by the North American mechanic, Murr [Moore]. In the soft clay that belongs to one of these systems of deposits, imprints of ferns can be discerned. They seem to belong to a neuropteris [nerve, sinew-like] which resembles the *Neuropteris acutifolia*. Other even less preserved impressions resemble gramineae [grasses].

According to Voznesenskii's specimen, a greenstone-porphyry with contents of pyrite stands in on the west side of Unga [Delarof] Harbor and on the east side of Sakharov [Zachary] Bay, along with other kinds of rock which evolved from the same through weathering or other influences. The basic mass of this material is greenish-grey, soft, and clay-like; and it contains in part strongly deteriorated crystals of feldspar and lightly interspersed chlorite; then diorite and a dense, not hard (2.5–3), but extraordinarily tough, dark aphanite, with albite crystals enclosed therein. A white crushable clay stone, perhaps the same blue clay with a high content of calcspar which is found 1 verst from the village (on the west side of the bay), resembles the well-known interactions of the diorite-porphyries with graywacke.

A silica-conglomerate to tuff (similar to that on Kodiak and Alaska) near Sakharov Bay, is extraordinarily rich in Tertiary bivalves of the most recent times. We have described them in the appendix, according to the specimens sent by Voznesenskii, Fischer, and Kupreianov. These are: *Cardium decoratum*, *C. groenlandicum*, *Venerupis petitii* var., *Pectunculus kaschewarowi*, *Saxicava ungana*, *Mya crassa*, *M. truncata*, *M. arenaria*, *Tellina lutea*, *T. edentula*, *Mytilus middendorffi*, *Ostrea longirostris* and *O. plicata*. The 40-fathom-high shore bank is supposed to consist of these deposits entirely. From the fissures of the rock cliffs of Sakharov Bay, which occur perhaps in amygdaloidal diorite, or in greenstone-porphyry, there is obtained, according to Voznesenskii's retinue: natrolite in bulby, spheroid, and kidney-shaped pieces with divergent, radiant cleavage; and then in rough, dense, radiantly composed pieces (completely analogous with the finds in the amygdaloids of Iceland); furthermore desmine, and stilbite in larger, thread-like pieces of milk-white to translucent color. The gouges consist of stilbite; the crystals on the inside of desmine (P. Naumann, Fig. 343). Finally in Fischer's specimens there are small pieces of an analcime crystal with yellow calcspar; and furthermore in all the specimens sent from Unga, from the same locality, large pieces of double-spar, grapecluster botryoidal-crystal-shaped stalactite-like chalcedony (the same as on Iceland), rock crystal and gem, crystallized quartz.

Among the boulders at Sakharovskaia Bay there were collected: milk- and orange-colored onyx, carnelian, and green jasper. In Voznesenskii's shipment there is also a sharp-edged piece of a yellowish quartz rock with inclusions of jasper, and furthermore soapstone-shale.

The collection of the Mineralogical Association in St. Petersburg has a piece of gneiss with a vein of quartz from Unga Island which contains molybdenite and copper; also limestone deposits of the type of the Karlsbad salt, and a diorite-amygdaloid with small, reddish analcime crystals.

It has long been known, that vegetable matter is silicified more frequently and faster on Unga Island than on any of the other islands. Thus, in higher elevations (especially on the island's northern side near Sakharovskaia Bay) there are petrified stumps and whole tree trunks. Some of these still show surfaces that have been worked on with iron hatchets (that is, during Russian times). Thus, they underwent that process within a hundred years. Just as with Kodiak, it could be surmised that the absorption of silica-soil depends on the kind of rock with which the wood fibers have come into contact. It is claimed that such contact-fossilization, or silicification by emanation, has been proven on Portland Island, where there is a stratum consisting of many tree trunks completely converted into masses of silica. This deposit lies on top of a corresponding deposit of an oolith group. But here, as on Kodiak, it remains perhaps to be demonstrated that the cause is simpler, more natural, namely that tree trunks, as well as deposits of rocks beneath them, were impregnated by silica-containing, albeit not necessarily hot, springs. In Voznesenskii's collection we have a specimen which is partially transformed into hornstone. From its characteristic pith ray it is easy to discern this as a kind of *Abies* [fir]. Another piece is metamorphosed into a lydian stone, which alternates with layers of pyrite. Opposite the NE point of Unga is hilly Popof Island. It has two lakes and four little rivers. A narrow promontory protrudes on the NW side, from where a sandbar extends toward Unga. East of Popof Island lies Korovinskoi [Korovin] Island. On either end of it there are two mountains standing alone. Between them are two lakes, from where two little rivers originate. South of Unga is the 28-mile-long island Nagai, which is narrow, mountainous and full of ravines. The mountains are transected by four valleys. There are lakes in two of them. There is supposed to be much clay-slate on the island. East of Nagai are the Koniuji Islands. The larger of them is mountainous, steep; and its mountains are transected by three valleys.

The fifteen to sixteen islands of the Simeonof group are almost completely unknown (cf. Veniaminov, I, pp. 263–265).

Veniaminov (I, pp. 244–254) groups the rest of the numerous islands between Unga and Unimak together under the name, Sannakh [Sanak] Islands (Aleut: Kyutkhin). They consist of the Pavlof group, the Belkof, and the Sannakh group.

The Pavlof Group

Veniaminov mentions five islands in this group:

1. **Peregrebnoi [Wosnesenski] Island** with a small bay, into which a small river flows, which originates in a lake. Its northern end is steep and consists of step-like rising rocks of basalt, and pillars.

2. **Ukolnoi Island.** On its southern side there are rocks from which a green color is obtained.

3. **Popereshny [Poperechnoi] Island** consists of a mountain of medium (?) elevation. On its SE side there is a cave.

4. **Dolgoi Island**, the largest of these islands, is little known. On its southern side a green coloring is collected from among the rocks.

5. **Goloi**, too, has not been much investigated.

The Belkof Group

This group consists of forty islands, the larger ones are: Sivuchii Rock, the two Yelasik [Iliasik] Islands, the Chishchelnoi [Sushilnoi] Islands and the Chernoburoi [Cherni] Islands. The composition of the rocks on them is unknown. The largest of all is Reindeer or Olenii [Deer] Island (Aleut: Animak), with a bay. This island is 6 versts long and highly elevated. On its NE side there are four peaks, which are cleft by deep ravines.

The Sanak Group

The island Sanak or Halibut (according to Cook) is 20 versts long and 3 to 5 versts wide, low in elevation, and rich in lakes. But in its middle there rises a cone, the so-called Halibut Head [Sanak Peak] (Cook, Vol. II, p. 117). Another mountain, not as high, rises from the western point of the island. Three miles east of Sanak is a narrow, rocky island, the rim of which is surrounded by submerged rock cliffs. The same is the case with another group of barren rocks west of Sanak. Thus, from three sides the island is unapproachable. Sanak has many small rivers, which all run southward. In the year 1790 (1788?) this island suffered a great innundation (Veniaminov, I, p. 27).

Sauer says (in Billings's expedition, p. 187): "In the middle of [Sanak Island] are three considerable mountains, joined together. The east and western extremities are low land. . . ."[55] [The entire island] is surrounded by a reef of rocks, some above water, and the surf breaking violently over others" (cf. also Zaikov in Pallas, *N.B.* III, p. 282).

Veniaminov counts the island Ikatok or Ikatan [Ikatan Peninsula] among the Sanak group as well. This island, however, adheres closely to Unimak Island. Therefore, it could with better reason be counted among the Fox Islands. Ikatok is separated from Unimak by a mere 200-fathom-wide strait. It is 15 versts long, 5 versts wide; and it consists of a mountain range of minor elevation. Off its northern and southern ends column-shaped rocks stand out to sea.

Before we turn farther westward to the Fox Islands, we shall mention a small volcano island, off the NW coast of the Alaska Peninsula.

Amak

It lies opposite Isenbeck Bay [Izembek Lagoon]. (Its southernmost point is at 55°25′ N. Lat., 163° W. Long.). It extends from NW to SE and is 4 miles long and 1.5 miles wide. According to Kastal'skii it is a dormant volcano which is covered from summit to base with rubble of different rocks changed by fire, as well as lava and pumice stone of different coloration. The shoreline is composed of volcanic boulders and enormous fragments of lava and basalt. More low-lying locations are covered with pitch sand. Amak is supposed to have had its genesis at the beginning of the last century. It was said to have had an active crater. But since Krusenstern (1804) it is supposed to lie dormant.

CHAPTER FIVE

➤➤ ◄◄

The Aleutian Islands

The Fox Islands
(Aleut: Kavalang; From Unimak to Amukta Island)

Unimak
(Lütke, *partie nautique*. Vues de l'isle Ounimak 8–14, et No. 17, and Cook, III, Vol. II, p. 410)

From Zaikov's report (Pallas, *N. B.* III, p. 281) we receive the first detailed news about Unimak, on which island he sojourned from 1775 to 1778. "The western promontory is rocky on both sides, and steep; and the shoreline is sandy, precipitous, and full of sandbars. The middle of the island is mountainous; and there is a volcano (Shishaldin), which is frequently on fire. On the northern side of the island there are two small rivers, one of which originates from a lake."

[Additional note from the original errata sheet, p. 422:] Unimak. On Helmsman Iakov's map (Coxe, German, p. 192) we first find (1769) Agneda Volcano (Agayedan or Shishaldin).[end of addition]

Cook (Vol. II, p. 117, [Eng. ed., London, 1784, vol. 2, p. 416)) sighted this volcano on the twenty-first of June 1778, and fixed its location at 54°48′ N. Lat. and 164°15′ W. Long. He believed, however, that it still belonged to the mainland. "Over this [Halibut Island, Sanak] and the adjoining islands we could see the main land covered with snow; but particularily, some hills, whose elevated tops were seen, towering above the clouds, to a most stupendous height. The most South Westerly of these hills was discovered to have a *volcano*, which continually threw up vast columns of black smoke." This obviously was Shishaldin Mountain of Unimak.

Khudakov's reports from his winter sojourn on Unimak, 1791–1792, are inconsequential, except for his map of the island (Lütke, *p.n.*, p. 291; Sarychev, II, p. 174).

In Sauer (Billings's exped. [Sauer, Eng. ed., London, 1802, p. 164]) it states on p. 197: "[The terrain of Unimak Island] is high, broken, and rugged, and there are three very conspicuous mountains upon it. The summit of the first [Pogromnaia?] is very irregular; the second [Shishaldin] is a perfect cone towering to an immense height, and discharging a considerable body of smoke from its summit [19 April 1790]; the third (Khaginak) has its summit apparently rent and broken, covered with snow, and towering above the fog which covered the middle of the land." Sarichev (II, pp. 28–29, with two profiles) sighted the three sugarloaf-shaped mountains on the eighteenth of June 1790. He named them Agayedan (Shishaldin), Khagman (Khaginak), and Kugidakh-Yagutcha (Pogromnaya). The first of them was smoking; the second had the appearance of a caved-in cone.

The most complete picture of this island to date, we have received from Lütke (*partie nautique*, pp. 291–298) and Veniaminov (I, pp. 204–220). Its length, according to Lütke, is 75 nautical miles, and its greatest width is 25 nautical miles, as it extends from SW to NE. On the whole, in form and constitution this island adheres closely to the Alaska Peninsula, from which it is separated by the Strait of Isannakh [Isanotski]. According to Postels, two parallel mountain ranges traverse the island. Lütke speaks of only one major range. But it seems that the Isannakh Range forms a second, independent ridge of mountains, sweeping from south to north on the NE side of Unimak, while the mountains along the NW coast form a side arm of the primary range, from which they are divided by a valley that runs lengthwise. The ridge that extends from Mt. Shishaldin to the SW end of the island, is transected by three or four deep valleys. The range sweeps closer to the south side of the island, but on the north side it falls off steeply. The northeastern end of the island is flat and covered with gravel. Farther toward SW, a flat, snaky coast extends along almost half the length of Unimak. From Cape Shishkov [Cave Point] to Cape Sarichef, and southwesterly beyond that point, the coast forms the base of Pogromnoi Volcano (54°30' N. Lat., 164°45' W. Long.), from which the just-mentioned promontories run out. The south side of Unimak is mostly steep and rocky, without any good anchoring places. And the SW end falls off in a steep rock wall, Cape Khitkhukh [Scotch Cap] (Lütke, *p.n.*, Profile 17). There are three villages on this island: Shishaldinskoi, Pogromnoi, and Nosovskoi. The way from the first to the second village leaves the mountains, which form the NW end of the island, on the right side, as it leads over an elevated plateau. Swampy terrain alternates here with mossy surfaces and rocky ground, until it reaches the coast again, north of Cape Shishkov. From Pogromnoi to Nosovskoi the way leads along the coast and along the base of Pogromnoi Volcano. Among the many rivers on Unimak, only three are of any importance: Red River (Krasnaia), which comes from a lake and flows NNE; Shishaldin River also comes from a lake and flows in a northerly direction; and Sand River (Peschanaia) streams westward. From among the numerous lakes, especially on the island's north side, one west of Shishaldin distinguishes itself by its large size. And another, small, lake stands out because of the sulfury taste of its water.

The island forms, as Lütke says, a veritable vault over a constantly active foundry furnace. The back of this vault is formed by a mountain range, which runs from SW to NE. Several funnels open into it, through which the submerged ocean of fire sends sparks and flames. These eruptions are so powerful that in spite of the large number of furnaces, the bottom of the forge often shakes. The largest fire chimney is Shishaldin Volcano, or Agayedan, as the natives call the mountain (also Sissagyuk according to Veniaminov). This volcano lies fairly well in the middle of the island, at 54°45′−48° N. Lat. and 163°59′ W. Long. It has a regular cone shape; and its elevation is reported by Kotzebue as 7,154 ft. and by Lütke as 8,953 ft. (1,400 *Tois.*). Chamisso believed that its snow-limit begins at 400 *Tois.* above sea level. But that might be erroneous, because he claims that in Unalaska the elevations ascend to 850 *Tois.* (cf. p. 69, comment). Shishaldin Volcano has burned since ancient times. But its activity is mostly limited to the belching of a great mass of smoke, which rises from the top straight upward. Thus Chamisso, for instance (Kotzebue, III, p. 165), saw the upper two-thirds of the mountain's massif clothed in snow. In some places it descended even lower toward the shoreline.

In the year 1824 and early in 1825 the eruptions of Shishaldin Volcano were especially violent (cf. Lütke's *p.n.*, or Berghaus: *Geography and Ethnography*, II, p. 742). And toward the middle of the month of March, a low ridge NE of this mountain split open in five or six places after a terrible subterranean detonation that was heard on Unalaska Island and on Alaska. Flames and black ashes were expelled, which covered the Alaska Peninsula all the way to Pavlof Bay. At high noon it was dark as night even in Morzhovoi Village, 10 German miles away. At the same time a flash-flood [*Joekulhlaup*] descended from the mountain, down the south side of the island, covering a section of land more than two German miles long, as it catapulted pumice stones along with it. But the flood did not last long. Even the waters of the ocean were murky far into the autumn season. Since that event the volcano has burned less strongly. The ridge, through which the subterranean powers released their pressure, continues to smoke constantly; likewise a small cone, which began to rise from the middle of the ridge. In November and December of 1830, Shishaldin roared terribly from out of the fog which enveloped it. After the fog had lifted, everyone was surprised about the black color which the mountain had taken on. The snow, which had always covered it, had disappeared; and long fissures, which expelled frightful flames, appeared on three sides simultaneously: on the north, west and south side. The northern side was constantly aflame. The fire erupted in spurts three times per minute and after every third or fourth normal emission there comes a stronger flame accompanied by sparks. In March of 1831, two fissures closed up. Only the northern one remained, which from base to top extended no less than one-fifth of the way up the entire elevation of the mountain. Its width is about one-seventh of its length. It looks like glowing iron, and it never changes its appearance. On the north-eastern base, too, the mountain is supposed to be on fire. After these eruptions, the natives believed they noticed a diminishing of the earthquakes. It is reported that blueberry bushes did appear in the aftermath of the falling of the fertile ashes near Pogromnoi village. Formerly they had not been there. In the aftermath of the eruption of 1827, the fish and shellfish near the same village became more rare. The former drifted about, dead on the ocean, and were washed ashore.

A little eastward of Shishaldin is a double-peaked volcano (54°45′–48′ Lat. and 163°45′ W. Long.). This is probably Sauer's Khaginak Volcano (cf. Lütke, *p.n.*, Profile 14), where the crater formation can perhaps be most readily observed. According to Veniaminov, this is the highest mountain on the island. In part it is connected with the mountain range at the NE side of Unimak (Isannakh Range). Since 1825, the latter mountain ridge smokes in five or more locations. But before that time there was one primary location where a mighty eruption had occurred prior to the diminishing of the activity of Shishaldin Volcano.

Another significant mountain summit is Progromnoi Volcano, mentioned above, and also called Nosovskoi, at 54°32′ N. Lat. and 164°42′ W. Long. It could, however, be that another, closer, mountain was indicated in this survey, which had been taken from the sloop *Moller* (cf. Lütke). According to Kotzebue (Vol. II, p. 1), it is a sugarloaf-shaped peak, 5,525 ft. high; somewhat inclining northwestward, as Lütke observed, and arising 6 miles from the SW coast. In 1795, there occurred a terrible eruption in this region. The mountain range broke open; and ever since that time the Pogromnoi Volcano has been at rest. But around it, volcanic activity was observed in numerous locations. Thus, according to legend, there arose in former ages a volcano on the mountain range that extends northeast from there. It has meanwhile collapsed. The same is said of a mountain which stands northwest of Pogromnoi Volcano. To this day some old people remember a small volcano on the north side of Pogromnoi. It expelled flames, but went dormant in the year 1795 when that range exploded with a ter-rible report, producing a thick rain of white ashes.

People furthermore remember another volcano near Cape Sarichef, west of the Pogromnoi Volcano. Presently only smoke rises from between large rock boulders. The water of brooks and swamps in that place is hot; and sulfur is col-lected there in large amounts. At the base of the now-dormant Pogromnoi Vol-cano, traces of earlier eruptions can also be observed throughout the region. A

5-verst-long distance can be observed half-way between the villages Pogromnoi and Nosovskoi, where burning has taken place for an extended time, because the rock looks as if it has come out of an iron smelter. People say that these rocks have come down together with the water which the mountain has given off(!). And in front of Nosovskoi Village ice was preserved for a long time after it had tumbled down, together with the water, in 1796. Farther southward, that is, toward the SW end of the island where heretofore no visible influence of the fire had been observed, there, according to Veniaminov, the mountains begin to swell up in one location. Lastly, on the southern side of Unimak, on the west side of Tugamak Gulf [Unimak Bight], the location of which is not accurately fixed, but probably not far from Pogromnoi Volcano, there, in September of 1827, according to the log of the corvette *Moller*, a "lave brulante" was observed, meaning most likely a not-yet-cooled lava stream. Dr. Isenbeck reports (oral report) that during the passage through the Strait of Unimak, a fiery column constantly ascended from the volcano closest to the coast. This was probably Pogromnoi Volcano, although Veniaminov's indications, (cf. below), according to which the Shishaldin erupted at the same time, speak against it.

It is questionable whether that cone in the island's interior close to the southern point, which burst apart in 1826, according to Veniaminov, is located closer to Pogromnoi Volcano or to the Shishaldin, or whether it is Pogromnoi itself. It is possible that the indicated point is close to the above-mentioned lava stream.

Among all the islands, Unimak is the most active theater of volcanic phenomena. As the result of them, the island endured the most drastic changes. The Russian clergyman Veniaminov has given the most complete description to date of these phenomena. Postels (III, p. 23), and Lütke (*p.n.*, p. 296) have made use of it. And then Baer has published it in excerpts in the *Beiträge*, Vol. I, pp. 173–176, and in Erman's *Archive*, Vol. II, 1842. But mainly the descriptions have appeared in Veniaminov's own description of the District of Unalaska, Vol. I, pp. 204–220. As a supplement to what has already been said, I give here Baer's translation of Veniaminov's manuscript, with additions from the Russian text, printed thereafter (I, pp. 205–207).

> On the NE side of Unimak Island, along the Strait of Isannakh, there sweeps across the entire width of the island a high, jagged mountain ridge, which has smoked profusely since 1825. I call it the Isannakh Range. West of that range there rise the two tallest, cone-shaped summits of the island. The first one of them (Khaginak?) is partially connected to the Isannakh Mountains. And it has a crater, which according to tradition developed long before (about 1690) the arrival of the Russians. If instead of this crater we imagine a continuation of the mountain, then it is quite obvious that this dome, rising in its present shape

above all others, must in its original form have been the tallest of all. It must have topped the others by one-quarter, measuring 12,500 ft. The second dome, called Shishaldin (Sissagyuk), lies almost in the middle of the island, but closer to the southern shore. It is completely separated from the rest of the mountains by low-lying terraces; and according to Lütke it measures 8,953 ft., Engl. The volcano has been continuously exposed to subterranean fire since times immemorial; and from time to time it erupts in flames. Thus it burned with a violent fire in the year 1824 and at the beginning of 1825, until March 10, that is, when the northeastern mountain ridge erupted. From then on until March of 1827 the dome only smoked; but thereafter it erupted in flames again until the year 1829. After this time the volcano smoked until autumn of the year 1830. Toward the end of that same year, however, it changed its appearance noticeably, and on its north side there appeared three fairly significant fissures, which extended from the rim of the crater far downward. Powerful flames shot from the gorge. Trembling and subterranean roaring were felt to alternate; and more than half of the permanent snow that had rested on the peak melted down from the northwestern and southern sides. Until March of 1831 the same phenomenon remained. Then the fissures closed up, one after the other. But when I traveled past that mountain on the sixth and seventh of May 1831, I saw it erupt with fire and sparks almost every ten to fifteen seconds, but not with equal power. From the crater on the northeastern side one could see a fissure descend, the length of which was one-fifth of the mountain, and its width was about one-eighth or one-tenth of its length. All the time the crack had the color of glowing iron; and it was transected horizontally by several isthmuses.[56] The foot of the mountain toward the NE was very hot and shaking noticeably. But beyond that, there were on this occasion no other eruptions, except for soot, which could be observed on top of the snow on the twentieth of April. In the autumn of the same year (1831) the mountain clothed itself again with snow. And since that time it only smokes.

In von Baer's translation (l.c., Veniaminov, I, pp. 35–37) it is further stated:

Far more violent here was the effect of the subterranean fire about the year 1795, and at the beginning of the year 1825. In 1795, during a violent west wind, the mountain range at the SW end of the island burst apart with a terrible explosion, erupting with an enormous quantity of ashes, white of color. Thus in midday the nearby villages, and even Unga Island, were wrapped in total darkness. The permanent ice which had covered this mountain range cascaded down on both sides, accompanied by many red-hot rocks. They came to rest halfway up the mountain and formed a wall or belt around it, which remains visible today. The place can also be discerned where the water flowed and where the ice came to rest for a few years after it had tumbled down. At this time it can be observed that in recent years the mountain is lifting or swelling up in one location. After that, there has not been any visible effect of the fire.

On the tenth of March 1825, the northeastern range of Unimak, which now smokes continuously, burst open in midday in five or more places with a violent report that sounded like a cannonade. It lasted almost the entire day and could be heard on Unalaska, Akun and on one end of Alaska. There followed an eruption of flames and a great mass of black ashes, which covered the entire point of the Alaska Peninsula. The entire region was covered with darkness. The masses of ice and snow that lay on the mountain melted and rushed downward for a time as a dreadful, five- to ten-verst-wide stream. On the east side of the island these waters descended in such large amounts that the ocean was murky in this area far into the autumn season, that is, long after it had abated. Thereafter Shishaldin Volcano, which lies not far from that range, ceased erupting with fire, as it had done until then; and from now on it merely smoked. In the middle of that range there emerged an elevation or a king of the hill, from which emanated clouds of smoke until the year 1831.

It should also be noted that the explosion and the thunderous din from the tenth of March 1825, could be heard on Unalaska. But on the SW end of Unimak itself it could not be heard. Therefore it must be assumed that the subterranean fire was directly connected with the Makushin dome in Unalaska. At the harbor of Unalaska it must have been very near the surface.

On the eleventh of October 1826, a dome on the island's interior, close to its northern point, burst open with a muffled roar and violent explosion of fire (bundles of flame, according to Lütke), emitting whitish ashes that covered parts of Alaska, Sanak with nearby islands (Chernoburoi [Cherni], for instance), and even Unga. It had been burning until the revolution of the southwestern shoreline (1795). Since that time there is a constant emission of smoke from the rock masses and very hot stones that had collapsed over each other. Burning sulfur can be discerned; and all the brooks and swamps in the region are so hot that they steam constantly. In August of the year 1830 the top of this dome split again, but without any extraordinary phenomena.

Between Unimak and Unalaska there is a group of islands called the Krenitzin Islands.[57] In shape and constitution they resemble Unalaska closely. The most important islands among them are Unalga, Akutan, Goloi (or Lentok) [Rootok], Avatanak, Tigalda, and Ugamak. These first three extend from SW to NE as extensions of Unalaska's mountain ranges. The rest form apparent spurs of that extension, which reach toward Unimak between Akutan and Akun from W to E toward NE.

Unalga has no high mountains. But it is a rocky island throughout with steep coastlines, which are lower on the western end (Lütke, *p.n.*, Profile 15). Between this island and the island of Akutan, six rock cliffs extend quite far above the water level. Their circumference varies between 700 and 100 fathoms.

The shape of **Akutan** is almost round; and the island is traversed by mountains in the most diverse directions. On the southern side they present a rugged aspect. In the middle of the island there rises at its highest point a volcano of 3332 ft. in elevation (54°08′ N. Lat., 165°54′ W. Long.). On top of it there is a small, deep lake; but on its northern side there is a crater, from which the Aleuts fetch sulfur. Close to the summit is a location that emits smoke; from time to time, according to Lütke; but according to Veniaminov, the emission is constant. According to Veniaminov, the place looks like an enormous whale, on the neck of which there is an opening where the smoke comes out. At a small bay on the NE side of the island, several hot springs gush out of the mountains.

Their temperature is so high, that people cook meat and fish in them. They flow into a little river, the water of which has normal temperature. On the SE side, three springs come from a rock wall, which exposes stratified deposits. Here is found obsidian in fragments, which are baked together on the surface. The southernmost promontory (54°03′ N. Lat.), which Veniaminov calls Cape Battery [Battery Point], is a branch of the volcanic cone that forms the right bank of Sarana Bay. There are parallel deposits here, which are traversed by vertical veins of a quartz-like rock. This promontory arises as a steep rock wall with low ledges along its sides. One of these has four embrasure-like caves. On the north coast of the island, too, a long, plateau-shaped cape juts out as a spur of the volcano. On the side facing the sea, there are several caves, and five archways (*vorota*). To the west of this point or on the NW base of the volcano, the remains of a mountain can clearly be discerned—a mountain that has collapsed into the sea. Now it forms a pointed hill of blackened rock. There is obsidian to be found along the shore. [The following addition is from the errata sheet on p. 422 in the original text:] From time to time (1760–68), sulfur flames are noticed on the volcano of Akutan (Coxe, German, p. 226) [end of addition].

Shelikhov (p. 145), and Cook (II, p. 124) do not mention these volcanic phenomena; but Sauer (p. 163) and Sarychev (III, p. 28) do mention them on June seventh, 1790 (cf. Schloezer's *Nachrichten*, p. 167).[58] Lütke and Veniaminov (II, pp. 192–195) provide most of the information concerning this island.

Akun. The mountains here are not high. One of them on the NW end (at about 165°33′ W. Long. and 54°17′ N. Lat.) has a pointed shape, and it constantly emits barely perceptible smoke (at intervals, according to Lütke). On this mountain top, sulfur is collected. In the region of the southern promontory of this island there are rock columns, grouped in the form of buildings. One of them stands out, because it has the shape of a human being. On this island, hard coal deposits are also supposed to occur (Lütke, Vol. III, p. 21). Off the west and southeast side of Akun there are four islets of 0.5 to 1 verst in circumference. On one of these, hot springs can be observed at low tide. Other than these islets, there are several rock columns standing in the sea. There is little we can report concerning the islands Goloi (the bald one, Ayaktak, Lentok [Rootok]) and Avatanak. From the latter we received deposits of quartzite with pieces of lydian stone baked in, from Voznesenskii.

Tigalda is high and rocky on its southern side. On its northern side it is flatter, and a low mountain range traverses its entire length. There is hard coal on its southwestern end (Veniaminov, I, p. 201) where a tall rock column rises out of the sea.

Ugamak is the name given two islands which rise abruptly from the sea to a considerable height. Only a narrow, barely perceptible strait divides them.

Unalaska (cf. Choris, *Voy. pittoresque*, Tab. VII and XI; Lütke, *p.n.*, Profile No. VII, X, XI, and XVI; Sarychev, II, Profiles to pp. 81, 82, View, p. 14, Maps, p. 150).

This largest and most often visited among the Fox Islands group of the Aleutian Islands is still very little known with regard to its natural history. Zagoskin (p. 14) even claims that its interior might possibly be unfamiliar to the natives themselves. According to Kotzebue (first voyage, vol. I, p. 166), a voyager will seldom behold as horrible and barren a spectre as this island presents, especially on its NE side. Black lava banks rise vertically from the sea and reach an elevation high enough to be covered with permanent ice. And the island seems to consist of nothing but peaks and closely crowded mountains, some of which are so tall that the summits reach into the clouds.

Krenitsyn provided us with the first reports on Unalaska (cf. Coxe, pp. 254–256; Pallas *N.B.*, vol. I, p. 256; Busching's *Mag.* XVI, p. 273), which had been discovered by promyshlenniks as early as 1760. According to his report, there are three bays on the northern side. One of these is called Udaga (Beaver Bay). It reaches NE to SW almost halfway into the island. The other bay is called Ignuk (Illuluk? Captain's Harbor). It lies southeastward and southwestward; and it can serve as an excellent anchoring place. There are two fire-belching mountains on this island (1768–69). One of them is called Aiagish (German Ajagisch) [Makushin]; and not far from there is a hot spring. [The following is a note from p. 422 of the errata sheet]: The Russians call the other mountain The Roaring Mountain. And from time to time (1760–1768) sulfur flames are observed on the mountains of Unalaska (Coxe, German, p. 200, 226) [end of addition].[59]

Everywhere the island is rocky, and the soil made up of clay.

Cook, who anchored in Samagundha Harbor [English Bay] in the fall of 1778 (Vol. II, pp. 121, 165, 186), observed absolutely no volcanic phenomena; and among the rocks on the beaches and in the mountains he noticed nothing unusual.

Sauer (in Billings's Expd., p. 303 [Sauer, Eng. ed., London 1802, p. 267]) stayed the winter in a small bay not far from Illuluk. He says:

> The soil is not deep, but black and fine, unmixed with clay or loam. It was with great difficulty that we procured, near the source of a rivulet, a sufficient quantity of clay to use as cement to our ovens, built with the stones collected on the seashore. . . . There are two extinguished volcanoes on this island (April 1792); and near one of these there was formerly a hot spring, but it is now buried under stones fallen from the mountain, which produces abundance of native sulphur. Earthquakes are frequent, and, by the account of the natives, sometimes very violent.

And in Sarychev (II, p. 13) is furthermore stated:

> On top of a mountain between Illuluk Harbor and Ugadaga, there are two small sweetwater lakes, the bottom of which consists of iron ocher, which can be found everywhere in moist places. On the seventh of June 1790, the Makushin Volcano was smoking (p. 132) [verified by Merck, *Siberia* . . ., p. 68]. But it has not burned for a long time. Only now and then (on Feb. 14, 1792) did it expel smoke. The mountain is somewhat flattened off; and southward it forms a ridge. On top of that ridge, sulfur and lava are collected. Earthquakes are rare. But in past times, they were extremely violent.[60]

Unalaska (in Aleut: Nagu-an-alaks/kha, that is: "That is Alakskha"), is 150 versts long and 50 versts wide. It divides into two unequal halves. The northern one is much more mountainous and wider; and many bays cut into it. The southern half is narrower and lower in elevation. There are three distinct mountain ranges on the island:

1. *The Makushin Range* consists of two parallel mountain ridges, divided only by a deep and narrow valley, and extending from SSE to NNW, from Makushin to Captain's Harbor. In the eastern range there is the highest mountain of the island, Makushin Volcano. Postels calls this range the Vessel of Mountains.

2. *The Bobrov (Beaver) Range* extends from SW to NE between Captain's Bay and Beaver Bay. Four more or less high passes lead across that range (Veniaminov, I, p. 159: "Eti gory peresekaiutsia chetyr'mia peresheikami raznoi vysoty, vozmozhnymi dlia perekhoda iz Kapitanskago v Bobrovoi zaliv").

3. *The Koshin Mountains* sweep through the remaining part of the island from SW to NE. The range is almost cut in two by the northern Koshin (Mokrovskoi) Bay [Pumicestone Bay], and the southern Kulylilak Bay. Thus the island takes on a choked or broken aspect. Beyond the two bays just named, toward the NE coast, the mountains get higher.

According to Postels, the summits of these mountains are sometimes rounded and sometimes pointed. The slopes are traversed by ravines, and on the N and NE sides they are very steep. The summits are completely bare, while their slopes and bases are thickly covered with brush, the green color of which contrasts sharply with the bald, often snow-covered heights. In the middle of August, 1827, snow covered 300 *Tois.* of the upper portion of Makushin Volcano (Lütke, p. 281).[61]

The volcanic cone of Makushin lies 10 *Lieues* [leagues?] north of the roadstead of Captain's Harbor,[62] and four *Lieues* distant from the sea itself (53°52' N. Lat., 166°48' W. Long.). It is 5,474 ft. (856 *Tois.*) high. Its shape is that of a completely preserved cone, only a little blunted at the top. On its blunted surface (a snow-covered plateau), there is a crater, containing sulfur, from which clouds of smoke are expulsed. On its southwestern side the mountain falls off abruptly and rises again to form two peaks, the eastern one of which is the higher one. It cannot be remembered, says Veniaminov (I, p. 160; see also below), that the Makushin Volcano ever spit fire. But occasionally subterranean rumblings can be heard; and in 1818 earth tremors were felt simultaneously. Thus it appeared to the people as if large alterations were occurring on the nearby island, Amaknak, at the time.[63] Langsdorff (II, p. 209) and Krusenstern (III, p. 142) report it to have erupted strongly with flames, while St. John Bogoslof Island, which had come into existence in 1795–1796, remained quite calm.

According to Eschscholtz (Kotzebue, III, p. 190), the steep and jagged rocks on the east and west sides of Captain's Harbor were subject to constant changes. "Where earlier travelers, like Sarychev, observed and drew cone-shaped peaks in 1790, there were now (July 1817) saddle-shaped indentations. What was formerly a peak, now covered the slopes as rubble. During his first sojourn on Unalaska, in 1816 (Sept. 6), Mr. Eschscholtz noticed that these depressions were flat. During his second visit, in 1817, they were already more hollowed out, and the low ledges of the side rim were converted into small peaks. Earthquakes were not the cause of such changes now, but most likely the uneven consistency of the rock-masses stratified on top of one another."

Veniaminov's observation (I, p. 178) should also be mentioned in this regard, which states that a lake in the vicinity of Veselov Village [Cheerful] (in the left corner of the bay of the same name), had formerly been connected to the sea. But now it is divided from it by a dry strip of land (*kosa*). No one knows when this took place; but there are unmistakable traces of wave formations and remains of dwelling places on the end of the lake, which faces the interior.

According to Postels, earthquakes and subterranean rumbling, which resemble cannon-reports, are quite frequent on Unalaska. They deceive the natives into believing they are hearing the signals of arriving ships and so they go to sea for their reception. This phenomenon is most frequent from October until April. In summertime it happens but rarely. In June of 1826 two strong earthquakes were felt, while Makushin erupted in flames.[64]

Most of the geologic material was brought from Unalaska by Chamisso (Kotzebue, first voyage, III, p. 165), Eschscholtz (Kotzebue, III, p. 192), Postels (Lütke, III, p. 17) and Voznesenskii. Hofmann, Fischer, Kupreanov, and Kashevarov brought but little.

The mountains in the interior, to the left of the valley, through which the road leads from the main settlement to Makushin, consist of fine-grained granite with black, isolated mica (Weiss and Chamisso in Kotzebue's voyages; Langsdorff, II, p. 29; and L. von Buch, *Canary Isl.*, p. 389). Closer to the volcano, and at the coast of the large bay, on the way toward Makushin, and near Makushin itself, there occur black porphyries (clay porphyry) rich in feldspar. They change into amygdaloid on one hand and on the other hand into greenstone, conglomerate-like porphyry, and true conglomerate. These types of mountains are stacked on top of each other in mighty layers, hardly sloped at all, which change seemingly without pattern. Only from a distance can the stratification be observed on the profile of these mountains. The porphyries have largely sharp-edged, jagged, needle-shaped forms. Only where they become conglomerate-like are their forms rounded off like granite. Several hot springs, 93° and 94° Fahrenheit, break forth from these porphyry mountains[65] near Imaglinskoe Bay [Summer Bay?], 5 versts from the main village, Illuluk [Iliuliuk; Unaslaska]. Their water is odorless and tasteless, and they contain deposits of limestone [*Kalksinter*].

Another not very hot spring wells up from a layer of true conglomerate at the foot of an insular hill of low elevation. This hill stands separately at the beach below the high-tide level. The deposits on top, which make up the content of the hill, are the common mixture of clay-porphyry.

West of Makushin Volcano (according to Postels), there is supposed to be a mountain which contains a soft, stratified rock that gets very hard when it is exposed to the air. Judging by some of the pieces that Postels received, it seems to be a tuff made up of lapilli. It contains small grains that resemble olivine, also augite crystals. The Aleuts use it to construct their baking ovens. And not far from the volcano, on the east side of a lake,[66] amber is broken from a rock wall. Voznesenskii sends Tertiary fossils from the northwestern base of Makushin, such as *Tellina lutea*, *Mya arenaria* var., *Venus*, *Turbo*, *Trocjus* in clay. Kastal'skii sends a *Triconicum anglicanum* (cf. Appendix I). And in Dr. Stein's treatise (*Trudy Mineralogicheskago Obshchestva*, St. Petersburg, 1830, pp. 382, 383) it is stated that someone, probably promyshlenniks, found mammoth tusks and molars on Unalaska in 1801.

On the west side of Captain's Bay the predominant rock is iron clay mixed with green earth and a hardened, ash-grey clay with small crystals of glass-like feldspar,[67] which takes on a porphyry-like appearance. And its slaty texture gives it similarity with porphyry-shale. This kind of rock occurs in deposits with clearly tablet-shaped stratification. It is for the most part strongly deteriorated and takes on a white color. In his editing of Eschscholtz's specimens, Engelhardt mentions grey-green porphyry from the same locality, covered with spots of green-earth and crystals of feldspar. On top of this lies a weathered, soil-like, greenish-grey porphyry.

On the NE side of Levashef Bay [Port Levashef], which rises to an elevation of over 2,000 ft., there are supposed to be significant amounts of granite-like syenite, which is transformed into gneiss. According to the statements of natives, the SW coast of Unalaska is composed of the same kind of rock, overlain with a shale-like rock (Lütke, III, p. 19).

According to Voznesenskii's index of rock formations, the following geologic conditions apply to the east coast of Captain's Bay:

In the southernmost corner of this bay there appear brown coal deposits; and farther in the island's interior toward Illuluk Village, there is claystone porphyry [saprolite], which seems to emerge from the granite. Near Illuluk Village itself there is a greyish quartz rock with pyrite, and yellow to green jasper-clay. On the mountain east of this village (according to Eschscholtz and Engelhardt) there is iron clay, dark and lavender-blue, with fine-grained, even cleavage. It tends to have a metal-tarnished surface. Above this are deposits of amygdaloid iron-clay. In it are elongated bubble chambers, some of which are empty; others are filled with green-earth and lithomarge [*Steinmark*]. There, too, is glassy feldspar in small crystals, also spheric excretions, the pieces covered with green-earth, which transmutes into apple-green lithomarge. Furthermore, amygdaloid just as described above, only without feldspar, but with pockets of dense hematite, enveloped in green-earth. These amygdaloids of Engelhardt's, along with other kinds of rocks on Unalaska Island, resemble the mountain formations of Unga (Sakharof [Zachary] Bay and Unga Bay [Delarof Harbor]). They could therefore be diorite-amygdaloids. Also east of Illuluk, but farther over toward the opposite coast of the island, there is a place where the rocks are placed together, crater-like. Voznesenskii collected a cryptocrystalline type of rock here. It is not hard, but tough, dense, and light green with bright-yellow particles. In this form its specific weight is 2.788; and it goes over then into a mixture of clay-slate and green-earth, with an oolitic structure, and a specific weight of 2.656. Both versions melt with difficulty along their edges into grey enamel. They might have originated from clay-slate. Farther northward from the location of these phenomena Voznesenskii found melaphyre standing between diorite-porphyry and labrador-porphyry, with a dense, green groundmass, in which crystals of labrador and albite can be discerned. In front of the blow-pipe it reacts the same as the metamorphosed clay-slate; but its specific weight is 2.710. At Cape Kalekhta (cf. Lütke, *p.n.*, Profile 16) there is an outcropping of brown-red porphyry, wherein augites are found. The so-called red wall west of Kalekhta is composed of this material. And between these points fieldstone and albite-porphyry alternate, reportedly with green jasper.

Furthermore Voznesenskii sends us from Captain's Harbor, without closer identification of the site, the following: gneiss and syenite (perhaps only deposits); diorite to clay-slate; claystone (with pockets of calcite), where the many augite-twins occur that Voznesenskii sends us, although (see below) they are also supposed to break in a basalt-like rock and in a porphyry-slate; also aphanite and crystals of amphibole. Genuine copper is found, according to Veniaminov (p. 171), at the banks of a lake, which is located behind Makushin Bay at the springs of a little river that comes from the north side of Makushin. Its banks are coated with ferric hydroxide.

Copper pyrite [chalcopyrite] appears in a rock wall on the north coast of Unalaska, 20 versts west of an abandoned village.

The following is mentioned as detritus in Kotzebue's voyage: hard, yellow clay with many crystals of pyrite; iron-clay with spheric excretions, almost completely covered with green-earth, porphyry-like, because of crystals of feldspar; and basalt-like rock from the foot of the Makushin Volcano. It is brownish-black, fine-grained, unmeltable, quite hard, and shiny at the cleavage. Augite is sprinkled in, and has porphyry-like crystals (Hauy's pyroxene hemitrope). There is porphyritic slate, greenish-grey with many glassy crystals of feldspar with raven-black and dark olive-green augite crystals, and grains of finely interspersed pyrite. There is brown-red porphyry with small crystals of dense feldspar, which is tinted by green-earth. Then porphyry: dark red-brown, dense, quartz-like dough with brick-red veins of jasper, green feldspar crystals and occasional almonds of zeolite. Amygdaloid: green-earth finely interspersed, calcspar infrequently in small almonds and pockets, glassy feldspar frequent, therefore on the whole porphyritic. Amygdaloid: the mass through weathering soil-like, greenish-grey. It contains almonds of calcspar coated with stilbite (Illuluk). Bluish-white, almost dense, quartz-like rock of feldspar with pockets of reddish-white porcelain-clay and abundant pyrite. Dense, liver-brown and bluish, black-spotted rock with fine, spear-like crystals of a brown-red, very soft undefinable fossil with finely interspersed feldspar in the first stages of weathering; straight-shelled separation. Furthermore: differently colored jasper (pyrite to jasper-clay), granules of mountain crystal, from the amygdaloid located east of Illuluk; carnelian, pyrite, obsidian. A small lava-bomb.

The following islands are situated close to Unalaska:

The island Biorka, Spirkin, or Sidanak [Sedanka] at Beaver Bay is separated from Unalaska by a strait "Ugadakh," very few fathoms in width. This island was first made known during Billings's expedition ([Sauer,] p. 185). According to Sauer it is 7 miles long and consists of barren, medium-high mountains. They consist of a hard, glassy, usually green, sometimes black, kind of rock (obsidian). The island's name is Sidanak (Sithanak), after a black, shiny, mineral color which is found there. On its eastern end there are two other small islets.

The island Amaknak lies in the middle of Captain's Harbor. It seemingly consists of four islets, the gaps between them having dried out. On the southern and northern ends there are two tall, conical mountains. Voznesenskii collected obsidian on Amaknak Island, and a lime-tuff with horn-corals. Postels (in Lütke's *Voyage*, III, p. 19) found on the east coast of this island much detritus of basalt with olivine, but no outcroppings of basalt. On the SW end rises a rock wall, a few hundred paces wide. On it can be observed many alternating layers of ferruginous clay, hornstone, jasper, clay-porphyry, porous basalt or basalt-like lava with zeolite and pockets of calcspar and adhesives which contain iron-clay. These layers are interspersed with vertical veins of a quartz-like rock. Farther eastward there follows upon these kinds of rock a porphyritic shale; and more southward there occurs a hardened iron-clay (Postels, l.c.). On the west side of Amaknak is a small, high, rock island, Uknadak [Hog].

Umnak

This island, made known by promyshlenniks in 1759, is separated from Unalaska by a strait 5 versts in width. In the middle of it, closer to Umnak, there is a peculiar cliff that looks like a ship under sail (Sauer, p. 260). At the southern opening of this strait, too, closer to Unalaska, a small rock island extends above the water. The island of Umnak, the length and breadth of which are 120 versts and 30 versts, respectively, resembles Unalaska much in location and shape. It begins southwesterly with a low point. Then its elevation and width increase up to about the island's middle, where Inanudak Bay from the north, and Deep Bay (Glubokaia) from the south approach each other closely, giving the island a bisected appearance. Between these bays there lies a low, hilly plain, which was mostly under water (Veniaminov, I, p. 140). But now it is covered with a bedload of sand and volcanic ashes. And on its borders, clearly visible, are steps in its bank. Northeastward from this plain, the island becomes once again wider and more elevated.

On the southwestern half of Umnak there is a mountain range. Toward Deep Bay it comes close to the south coast; and there are two volcanos, Recheshnoi Volcano (53° N. Lat., 168°24′ W. Long.) and Sevidovskii [Vsevidof] Volcano (52°10′ N. Lat., 168°12′ W. Long.). But then this range merges with a high mountain range, which sweeps straight across the island at right angles to the other mountains. Mt. Vsevidof is the highest mountain on the island. But it does not reach the elevation of Mt. Makushin of Unalaska. Shelikhov was the first to mention it (p. 144, July 1784) as a smoking cone, with hot springs emerging from the foot of it. Sarychev saw it (II, p. 7) in 1790 on the thirtieth of May. It was smoking and covered with permanent snow (Profile appendix, pp. 81 and 86). And Veniaminov (II, p. 139) reports that its summit is ridge-shaped. I find Recheshnoi Mountain first mentioned by Veniaminov (l.c.). At one time it was also to have "burned."

On the northeastern half of the island, the mountain range extends analogously to the southwestern half. At the eastern side of the plain, between Deep Bay and Inanudakh Bay, there is a range of hills running from SSW to NNE, parallel with the one on the western side. But connected to this is a range which has a northeasterly direction. It ends 10 miles west of the high, abruptly descending Cape Tulikskoi at the pointed volcano-cone Tulikskoi [Tulik] (53°20′ N. Lat., 167°50′ W. Long.). That, in turn, connects to a mountain chain with northerly direction.

So far as the condition of the island's coast is concerned, it is not high in the south, but steep; and there are several rivers which have their source at Mt. Sevidof (Lütke writes Vsevidof). Near that coast are the Vsevidof Islands. On the west side of Deep Bay, into which the largest river of the island flows, the coast reaches an elevation of 15 *Tois*. Then follows a low, sandy coast. But opposite Unalaska it is rugged, all the way to Kettle Cape (Mys Kotel'noi), which falls off absolutely perpendicular as seen from the seaside. But toward the land side it becomes low and merges with little hills into the northeastward-extending mountain range. It can be discerned at first glance (Veniaminov, I, p. 32), that this promontory is what remains of a once mighty mountain. About the cut-up, rocky coast opposite Unalaska, Lütke also states (p. 299) that it seems to be the remnant or the base of mountains which collapsed into the sea. Sandbars and rock columns are frequent along the entire south coast. From Cape Tulikskoi to Cape Yegorkovskoi [Tanak], and southwesterly from the latter toward the width of Tulikskoi Volcano, the shoreline is sometimes low, sometimes high, without bays, but often transected by ravines and creeks. But then it becomes mountainous and rocky toward the main settlement, Recheshnoe [Nikolski], which is located toward the SW end of the island, surrounded by lakes and swamps.

After Unimak, Umnak Island seems to be the most lively theater of volcanic activity in historical times. The island's northern side is especially significant in this regard.

In the year 1817 (on the second of March, according to Lütke's *p.n.*, p. 300, and Veniaminov, p. 37; or on the first of March, 1820, according to Postels in Lütke, III, p. 24) the ridge of a mountain on the northern end of Umnak (Tulikskoi?) burst open and threw out ashes and rocks. The former reached Unalaska and even Unimak. The latter were carried a distance of 5 versts. A strong earthquake, accompanied by a horrible SW storm, transported the inhabitants of Unalaska into the greatest fear. Near daybreak it was found that the ground was covered with ashes from one to several feet deep. Even the brook that ran along the foot of the establishment on Unalaska was filled with ashes and had no fish during the entire year.

Not far from the place where the eruption occurred, the Aleuts now collect amber, which is found in a removable soil, which covers a slope. A lake touches the foot of that slope. The Aleuts have two boats on it, which they cover up with a fur. Onto that fur they throw the soil, from which they glean the amber.

There is a mountain ridge on the northeast side of the island, according to Lütke. Postels mistakenly placed it on the northwest coast. It opened up in the year 1824 and was still emitting columns of smoke in 1830. There is also a small cinder cone (Recheshnoi?) on the SW end of the island, which became active in August of 1830.

The village Egorkovskoi, now abandoned, located on the southwestern half of Umnak at Deep Bay, was once located on the northeastern end. During the eruption of 1817, while fortunately the inhabitants were away on the Pribilof Islands, the village was buried under enormous rocks and ashes. The place where the village is presently located, is supposed to have been under water once, according to legend. After the earthquake it is supposed to have risen up.

Proof of the constant activity of subterranean fire is the many hot springs on the island. They are especially numerous in a little valley in the mountains NE of the Vsevidof Volcano.[68] One of them is conspicuous because of its similarity with an Icelandic geyser, by its periodic emergence and disappearance. The spring, which is so hot that meat and fish can be boiled in it, spouts its water two feet upward, four times per hour. It disappears just as often, without leaving an opening behind. The ground consists of gravel and sand that can be walked on. Just before the geyser erupts, noise can be heard from underground. Since 1828, several new springs have reportedly emerged. On the southern slope of the Vsevidof Volcano, not far from the abandoned Egorkovskoi village, in a NE direction, 1.5 versts from the coast, there are three springs in a small valley. The middle one of these is hot, too hot to immerse one's hand in. Another (a mere 4 *arshin* away from the first) is cooler, and the third is a normal temperature. The Aleuts bathe in these springs, and they claim that previously the one in the middle was cold, and one of the others hot. They have no sulfuric odor. On the north side of Mt. Vsevidof, at Inanudak Bay,[69] there is a spring so hot that fish can be boiled in it.

Traces of volcanic activity are found on the entire island. There are, for instance, rocks scarred by fire; they lie about exposed or covered with a thin layer of soil. On the southwestern end this is least noticeable. On the promontory opposite Samalga Island, coral (reefs?) are found, according to Veniaminov (I, p. 150). On the west coast there are three rocks, which belong to the base of Recheshnoi Volcano. One of these is composed entirely of granite. The west coast of Inanudak Bay consists of the same material. On top of these granite cliffs, mighty boulders can be observed, which had been exposed to the effects of fire. On the N and NW coast, and north of the oldest settlement, Egorkovskoi, black and green obsidian is found, which the Aleuts call arrowheads (Russian: *strobuchnoi* [*stravochnoi?*]). These obsidian-boulders reach a weight of 100 pounds. The same is true of the trachyte and porphyry boulders which can be seen lying about. They had been expelled in 1817 during the eruption of Mt. Tulikskoi. Not far from here a clay-like rock is found which is not affected by fire. Therefore, it is used for the building of ovens. Eschscholtz and Chamisso (p. 166) obtained petrified wood, fragments of large trunks of dicotyledons, which supposedly originate from the bottom of a lake on Umnak Island, which dried out as the result of an earthquake. In the same place it is also mentioned that rocks, expelled in recent times (before 1817) from the island's volcanos, had filled up a channel, which had been navigable before then.

Now we address a geologic phenomenon, a part of the volcanic activities on the northern side of Umnak, about which L. von Buch, our famous geologist, says: "few might be more instructive, and more universally applicable." We speak of the island:

St. John Bogoslof or Agashagokh

The first report about it we received from Krusenstern (III, p. 142) and Langsdorff (4° II, pp. 208–211; 8° II, pp. 324–328); and from Khvostov and Davydov (II, p. 176), Lisianskii (II, pp. 43, 135); Kotzebue (II, p. 106; III, p. 166), Chamisso, Vasil'ev, Dr. Stein (*Iziasnenye risunka iredst. vulkanicheskii ostrov Sv. I. Bogoslov*, St. Petersburg, 1825), Baranov and Teben'kov (Lütke, *p.n.*, p. 302).

According to Baranov's report (cf. Krusenstern in *Mem. Hydr.* 1827, p. 97, with the map of St. Bogoslof in Krusenstern's *Atlas*, Plate 19, 1827, or in Berghaus, *Geography and Ethnography*, II, p. 738) there suddenly arose a storm from the north on the first day of May 1796; and the sky grew dark, which lasted the whole day. During the following night the storm increased. On this and the following day a muffled din could be heard, and a far-off crashing sound, which resembled thunder. At the break of the third day the storm abated and the sky cleared. And now a flame was observed between Unalaska and Umnak, and north of the last-named island. It arose from the sea. And soon there was smoke, which lasted ten consecutive days. After that time something white, of rounded shape, was observed rising above the surface of the sea; and it grew very rapidly in size. This is the way it appeared from Unalaska; while Kotzebue (cf. also Buch's *Canary Isl.*, p. 387) saw the same from Umnak, and described it as follows:

On the 7th (18th New Style) of May 1796, an agent of the Russian American Company, Mr. Kriukov, was on the northernmost point of Umnak.[70] Storm from NW had obscured the view toward the sea. On the 8th, the sky cleared up; and now a column of smoke could be observed rising from the sea a few miles off shore. Toward evening there was something black, which arose beneath the column of smoke just a little above the surface of the water. During the night, fire spouted upward from that location. At times this was so strong, and the amount so large, that on this 10-mile-distant island (25 versts, according to Veniaminov, I, p. 156) all objects could easily be discerned. Now an earthquake shook the island, and a horrible din echoed back from the mountains in the south. The emerging island threw boulders all the way to Umnak. With sunrise the earthquakes abated, the fire diminished. And now the newly emerged island could be seen in the shape of a pointed hat. One month later Mr. Kriukov found the island significantly taller. It had in the meantime continued to expel fire. Since then it has continued to grow in circumference and elevation; but the flames have abated, and only steam and smoke remain to be seen continuously.[71]

Four years later no more smoke was seen; and eight years later (1804) hunters visited the island. They found that the water was warm; and the ground was still so hot, that in many places it could not be walked on. Even a long time thereafter the island continued to grow in circumference and size. A Russian of very healthy judgement reported that the circumference amounted to 2.5 miles; the elevation to 350 feet. In the surrounding 3 miles the sea is covered with rocks. From the middle of the island to the point, he found that it was warm. And the steam, which arose from the crater, had a pleasant odor, probably from the mountain oil. A few hundred fathoms north of the island there arises a rock column of considerable elevation, which Cook called Ship Rock in 1778. Cook, and after him (in 1790) Sarychev, sailed through between this ship rock and Umnak with full sail.[72] The elevation of St. J. Bogoslof is probably judged too low, according to Buch. At such a circumference the elevation should easily have been several thousand feet. Langsdorff's expression points up the same thing, when from his own perspective he calls the elevation a medium one. When he got his first view of the island on August 18, 1806, there were four conical mountains visible on the NW side. They rose by steps to a medium and highest elevation. The latter seemed to rise on all sides vertically like a column (Langsdorff, II, p. 209). In April of 1806 the island had been visited from Unalaska. From the northernmost point of Unalaska its location is due west, 45 versts distant. According to Teben'kov (Lütke, p. 302), 42 Italian miles from Cape Veselovskii. Rowing around it took six hours. Climbing the peak, straight up from the shore, took a little over five hours. On its northern side it was burning; and lava, a soft material, flowed from the summit down to the sea. In the south the ground was cold and more flat. On the slope there appeared many hollows and fissures which expelled masses of steam that left sediments of sulfur. At that time (1806) it was still visible how each year the island grew in circumference and the summit in elevation. Chamisso (in Kotzebue, III, p. 166), mentions that vegetation had begun to cover the island, according to reports he received.

In Baranov's narrative, which erroneously reports the time of eruption ten years too late, the following events are mentioned (cf. above); although it is not certain that 1814 should indeed be read 1804 in what follows:[73]

On the first of June 1814, a *baidarka* was launched to observe the phenomenon from a closer proximity. When its occupants had approached to a distance of five versts, a violent current was observed between the pointed blind cliffs. In spite of this it was possible to go ashore at a very low point, where sea lions had occupied the rocks in large numbers. It appeared that the island consists entirely of precipices, covered with small rocks, which are continuously expelled from the crater. They obscure the view and nearly cover the entire surface of the island. Therefore it was impossible to make investigations on land. It was instead decided to sail around it. Nowhere could sweet water be found. In the year 1815 (1805?) a second expedition was sent to the island. This time the island was found to be much lower in elevation than in the previous year. The bad weather forced the people to remain there for six days. A very strong current flowed continuously around the island. The physiognomy of the island had changed completely. There were ravines filled with masses of rock, which continuously collapsed, whereupon new precipices opened up.

The next, more detailed report we find in Stein's remarks, mentioned above, to the (rather sketchy) illustration of the island St. J. Bogoslof (cf. also Lütke, *p.n.*, Profile No. 18). We hear that Alexei Petrovich Lazarev, second lieutenant on the imperial sloop *Blagonamerennyi* under the commanded of Lieutenant Commander Gleb Semenovich Shishmarev, was unable to land on the island because of the strong surf. In the boat, which made the attempt on the second of June 1820, there was also Dr. Stein, to whom we will now give the word.

During our circumnavigation of the island we saw on its southern end (Cape Sarichef) a large number of sea lions (*Phoca jubata*). And from the highest point of the mountain—which I call Krusenstern Volcano—there arose, probably from the crater, columns of smoke (not fire). From a crevice at the foot of the mountain, a spring spouted up in an arch like a waterfall. The entire island is a bare and barren rock; and only in a few places was there still snow, covered, it seemed, with volcanic ash. On the sketch there appears on the right side Cook's "Ship Rock," which is much favored by the birds. And the island St. J. Bogoslof presents the spectre of deep fissures and lava streams (*ispeshchren*), traversing its surface. Its circumference amounted to 4 Italian miles (7 versts), and the elevation 500 ft., Engl., above sea level.[74]

From this information we discern that the report of that Russian mentioned by Kotzebue is perhaps not as erroneous as Buch assumes, and that, Langsdorff's data notwithstanding, the circumference of the island had almost doubled from 1804 to 1820, while the elevation, too, had gained very gradually by 150 ft. Veniaminov merely remarks (I, p. 34) that the island has not grown since 1823.

According to Teben'kov (in Lütke's *p. n.*, p. 302), its location is 53°58′ N. Lat.; according to Vasil'ev 53°56′20″ N. Lat. and 167°57′ W. Long. In 1832 it had no more than 2 (Italian) miles of circumference, and an elevation of 1,500 ft. Its shape is that of a pyramid, the sides of which are covered with pointed rocks, which threatened to collapse at any moment. The north coast of the island is rugged. The southern coast is a steep wall. From the latter side a low spit of land extends into the sea. This is where sea lions are resting. One verst farther north of this point a rock[75] extends, quite high, above the water level; it probably has no closer connection to St. J. Bogoslof. Teben'kov sailed through between the rock and the island before a full wind, without noticing any hidden cliffs, which had been mentioned earlier by Krusenstern regarding this place.

Thus according to these reports the island diminished in circumference by almost one half, but grew in elevation by 1,000 ft. The eruption and earthquake that afflicted Umnak in 1820 and 1824, could help explain this enormous growth taking place in such a short time. More plausible, however, would be an error in Mr. Teben'kov's observation.

If we compile the different reports, then the island went through the following dimensions since it emerged in 1796:[76]

1804 (Kotzebue) 4.3 versts (2.5 nautical miles) in circumference and 350 ft. in elevation.

1806 (Langsdorff) 10–15 versts circumference, and 2,500 ft. elevation.

1815 (Baranov) it became lower.

1819 or 1820 (Vasil'ev) 7 versts circumference and 2,235 ft. elevation.

1820 (Dr. Stein) 7 versts circumference and 500 ft. elevation.

1823 (Veniaminov) the island stops growing in circumference.

1832 (Teben'kov) 4 versts circumference, and 1500 ft. elevation.

It is not possible to bring the data into complete agreement. Still, from them seems to emerge the fact that in 1814 (the year closest to the report of Baranov) St. J. Bogoslof had the highest elevation. It is furthermore not impossible that Langsdorff and Vasil'ev calculated the elevation too high, and Stein the same too low. Teben'kov's calculation might have been too high as well. For the five-hour-long climbing of the mountain on the part of Langsdorff, a minimum of 2,500 ft. was assumed. But if the island grew in elevation until 1815, it could well have reached an elevation of 5,000 ft. The assumption of a repeated rising and falling is too presumptuous; it does not serve to unify the data. We can assume with much probability that other islands, too, in the region of Umnak, such as the Sevidovsky [Vsevidof] Islands, Tanghinakh [Pustoi], Samalga, have resulted from similar events not too long ago. I also feel that I ought to mention a report in this place, which is mentioned in *Khronologicheskaia istoriia otkrytiia Aleutskikh ostrovov*, pp. 119–128, although the author, Vasilii Berkh, does not point to a similar possibility of a volcanic phenomenon.

The year 1798 was the last time a privately owned ship, the *Zosima i Savvatii*, was outfitted. It was owned by the Irkutsk merchant Kiselev. After departing from Okhotsk, she passed the second Kuril Strait and sailed northward along the coast of Kamchatka to Cape Kronotskii. And thereafter she sailed easterly to Bering Island and Copper Island, where she stayed for three years, and where the hunting was plentiful. After that time the Near Islands were visited; and here a conference was held in which direction to proceed. The reports reveal that the skipper was not at all equal to his task. Hardly any credit can be given to the indication of directions taken by the ship. Supposedly they took a northeasterly course, and then a more southern one. But it rather seems as if the *Zosima i Savvatii* took first an easterly, and then a southeasterly direction. We shall let the first sailor of the ship have the word, whose trustworthiness Berkh does not doubt.

> It was our desire to visit islands located farther away. Therefore we held a conference regarding which course we ought to take. But since the opinions of the promyshlenniks, who had made several voyages before that time, were divided, our "boatswain" was consulted (who had been assigned to us by the chief of Okhotsk Harbor). At this occasion we were again convinced that he knew nothing (*chto on nichego ne znaet*). Finally, after much talking back and forth, a decision was made to steer NE for a few days, and then straight southward.

Since the Aleutian Islands form a chain, as our experienced sailors maintained, then we should in any case come upon an island, where we would find out what next to do. So we decided to follow that plan, holding straight to NE. How long we followed that course I cannot tell exactly. But I do know that hereafter we steered southward with favorable wind. During the first six days of our voyage we did not worry. But when during the next two days there still appeared no land, the crew began to grumble. Several of them thought we surely had already passed the Aleutian Islands; because of their small distance from Bering Island we ought to have reached them with such a good wind. The grumblers were appeased by the older more experienced sailors and by the captain (boatman). The former said: you don't understand a thing about this matter. The Aleutian Islands are so close to each other that one has to jump over them if he intended to avoid touching them. And the boatman assured us that he probably took his course farther northwards than necessary, since he did not know how to make the calculations. We allowed ourselves to be persuaded by such reasonable arguments; and so we continued our southward journey. But how great was our astonishment, when we began to feel warm air; and here it was the middle of October (*nachali oshchushchat' teplotvornyi vozdukh*). A few days later, that is, already in November, the heat reached such a degree, that the pitch which covered our vessel began to melt. This new incident caused the crew concern. And after we had argued a whole day (12 hours), thinking about what to do, we suddenly saw an island[77] in front of us, and surrounding the ship many young fur seals (*Phoca ursina, morskie kotiki*). Instead of concerning ourselves with this new item, and making use of this new discovery, we based our decision on the counsel of the elders, to concern ourselves neither with the island, nor with the animals. They were perceived to be an apparition of the evil one (*iako nechistago privideniia*). We should instead steer to where the wind blows. Thereupon the all-wise Providence brought us a vigorous wind from the south; and our argonauts resolved to steer northward.

They held this course for twelve days and reached the island Afognak (NE of Kodiak), where they found a settlement of the Russian American Company. Here Baranov assigned to them one of his own seamen, who brought them to Unalaska. They reached Okhotsk in 1803.

G. I. Davydov reports that course of events differently (Khvostov and Davydov's *Voyage, 1802–1804*, Russian, I, p. 158). He says that only during the return voyage was the island in question sighted, after they had sailed with the wind for eleven days. Therefore, the point is to be sought below 40° N. Lat. V. Berkh considers the report of the first sailor of the *Zosima i Savvatii* as more correct. He calculates the turning point of the voyage, and the sighting of the island, to be between 43°30' and 44° N. Lat. and 160° to 165° W. Long. In this region Portlock and Dixon observed a large number of fur seals and one sea lion (who do not depart more than 25 versts from shore). But since during the time of their presence, i.e., in April, May, and June, very strong fogs are prevalent there, an island could not be sighted. Lisianskii's ship, the *Neva*, sighted a sea otter on the thirtieth of June, 1804, under 42°18' N. Lat., and 163°W. Long. But the horizon was wrapped in darkness. During her return voyage the *Neva* observed a large number of fur seals (*kotiki*) at 48° N. Lat., and 140° W. Long. And Veniaminov mentions (I, p. 512) that according to the Helmsman Petrov, there was an island in the region, extending from W to E. It is no more than ten versts long and had volcanos (*sopki*) on both ends.

In our opinion V. Berkh has calculated the latitude too far south. One is forced to assume a straight line to support the main point, namely the twelve-day journey from the new island northward to Afognak Island. Berkh, too, does not assume this, inasmuch as he positions the longitude between 160° and 165°. It follows that the distance, calculated at three knots (5.25 versts) per hour, is 15 degrees or 900 Italian miles (1,666 versts) too large. Two knots per hour, or 3.5 versts, might come closer to the truth. For twelve days this amounts to 1,000 versts or not even 10 equatorial degrees. If we assume Berkh's indication of longitude as the mean, then we arrive at a point no farther south of Umnak or Unalaska than St. Paul Island (Pribilof Group) lies northward.[78]

It is not impossible that the appearance and disappearance of islands in this ocean has taken place more often than is known to us. And the uneasiness of the otherwise so sagacious Russian seafarers of the above-mentioned narrative seems sufficient to indicate that this was an unusual natural phenomenon confronting them—perhaps an island just recently emerged.

The Islands of the Four Mountains
(Ostrova Chetyrekh Konochny [Sopochnyia])
(Choris *Voy. pittoresque.* Views on Tab. II)

In the chain of the Fox Islands, these form a special group. They perhaps best characterize, according to Buch, the formation of the Aleutian Islands. There are five of them, or six, counting Adugakh Island. Pallas mentions them by the following names (in Busching's *Mag.* XVI, p. 273): Kikalgist (Kigalga or -gan), Kagamila (Kigamiliakh or -yek), Tana-Unok (Tanakh-Angunakh), Chigulak (Chegulakh or Chagulek), Ulaga (Ulliagin or Ulaegan). Sarychev (II, pp. 6–7, with Profiles in which the tops of the cones are shrouded with mist) calls them: Chagamil (Kigamilgakh), Tana (Tanakh-Angunakh), Chiginok (Chegulakh),

Uliaga (Ullyaegin). The group is called Unugun in Aleut (cf. Veniaminov, I, pp. 134–138).

Kigalgan [Uliaga Island] is a cone, eruptions of which have not become known. Its northern side is steep and subject to landslides, which seem to indicate the presence of vent holes.

Kigamiliakh [Kagamil Island] (52°53′ N. Lat.) is of elongated form with not very tall mountains. A few of them are supposed to have once erupted. On the southern side in mid-elevation sulfur is found that is still quite warm. The ground is warm and emits steam. Subterranean rumbling can also be heard at this point, and hot springs emerge from the foot of the rocks. A cave is known on the west side of the island.

Tanakh-Angunakh [Chuginadak Island] (53°N. Lat., 169°45′ W. Long.) is the tallest island of this group. Its shape is oblong with a steep south side. An active volcano rises on its western side. According to an Aleutian legend it once formed an island by itself. But the dividing strait was filled in when a mountain collapsed. At the foot of this mountain there is a spring so hot that it can be used to cook in (Busching's *Mag.* XVI, p. 273; and Veniaminov, I, p. 137).

Ulaegan [Carlisle] and **Chegulakh** [Herbert] are rugged volcanic cones without brooks or bays. At the peak of the latter there is a kind of crater. Both were still active in the eighteenth century, according to reports (traditions according to Voznesenskii). Between this group and Umnak Island lies the small, low, island Adugak. It is covered with many fills, has no lakes, bays, or rivers. The Aleuts report that on this island there is a freshwater spring, which takes on a bitter, stale taste before the onset of a storm. It emerges from a mountain, 10 *Tois.* above sea level. Voznesenskii sends us from this island much, more or less deteriorated, clay-porphyry (trachyte), white of color, as it occurs also on Unalaska. In a similar lode, reportedly a sediment of hot springs, no infusoria are contained.

[Commentary on the Four-Cone-Islands from the original edition, p. 422:] According to Helmsman Iakov's map (Coxe, p. 192) there are volcanos on the islands Kitalga, Kagamila, Ulaga, and Chagulakh, which were sighted on July 4, 1769 [end of commentary].

The three westernmost of the Fox Islands, which follow the Islands of the Four Mountains, are considered by Veniaminov as belonging to the former group; and they are not without interest.

Yunaska is mountainous, but not as high as Unalaska or Umnak (Lütke, *p.n.*, View 19). On the east side (52°40′ N. Lat., 170°28′ W. Long.) there is a volcanic cone, the eruption of which reportedly occurred in 1824 (according to Veniaminov, I, p. 38, in the year 1825). It was accompanied by expulsions of rocks until June. Choris, however, saw it smoking steadily as early as the fifth of April 1817 (*Voy. pittoresque, Iles Aleutiennes*, p. l, with view on Tab. I). In 1830 its summit erupted in flames and ashes, which colored the descending snow or rainwater, and even the ocean water near the coast, black. There are no hot springs, brooks, or bays.

Chugul (Chuginak [Chagulak]) is actually a mere rock column, 3 miles in circumference, the elevation of which reaches that of Yunaska and the following islands. Its location is approximately 52°36′ N. Lat., 170°56′ W. Long. It consists of rocks that threaten to collapse. And only birds and sea lions live there (cf. Sarychev, Vol. III, p. 6, view of the island; and Lütke, *p.n.*, Profile 20).

Amukta. The interior of the island (52° N. Lat., 171°04′ W. Long.) is mountainous. The highest summits are irregularly shaped. According to Lütke (*p.n.*, Profile 20) they are cones with depressed peaks. The coasts are low, but rugged. Not far off its southern end a tall rock column extends above the water. Here, too, there are neither rivers nor bays. Volcanic phenomena had once been apparent (cf. Schloezer's *Nachricten*, p. 107; and Busching's *Mag.* XVI, p. 273: "On the island Amukta there is a fire-belching mountain"). And Shelikhov says, p. 56: "Amukta seems to be engulfed in flames (June of 1786) from the volcanic mountains." Sarychev (II, p. 6 with profile) indicates (May 1790) that there is still a fire-belching mountain there. But by 1830 everything was calm, because Lütke and Veniaminov do not mention any volcanic phenomena.

The Andreanof Islands (Negho)

The first island of this group, which follows Amukta Island, is **Signam**, or **Segnam** [Seguam]. A mountain range, transected in three places, traverses this island; and in a few locations it is covered with permanent snow. The NE side is higher than the NW side. And Sarychev remarks in May, 1790 (II, p. 6 and p. 178 profile) that one of its mountains has once belched fire. At the eastern end there is a small volcanic peak (52°30′ N. Lat., 172°12′ W. Long.). Its color is black; and from time to time it emits a thick, black smoke, which then disappears again. This smoke rises occasionally from two or three locations of the central mass of the range. On the NE side of the mountains it seems to arise vertically from the water. The N and NW coasts in contrast rise in green inclines to the mountain ridges. The southern coast is low, but rough. There are many hot springs here, and holes which expel mud, such as are described in more detail in Atka.

From Signam Island Mr. Blaschke sent me lapilli, a large amount of olivine grains (crystalline in part), and augite. Furthermore obsidian, sulfur, graphite, quartz, calcspar and glassy feldspar.

Amlae [Amlia] Island is long and narrow, extending from W to E. Its middle is occupied by a range of mountains. Most of them are cone-shaped hills of medium height (compared with those mentioned earlier). No active volcano is known there; and in its vicinity there are relatively few shoals and no islands. According to A. Tolstykh (Busching's *Mag.* XVI) there are many mountains on this island, where several small rivers have their source. And according to Sarychev's indications (II, p. 6 with three profiles, and p. 178) it seems not to have changed since May of 1790. He says its eastern end consists of a high, steep rock wall (Lütke, *p. n.*, Profiles 21 and 22). From the northern coast of its west end Voznesenskii broke trachydolerite, which predominates on Amchitka and Bering Islands. But mica was more in evidence here; hornblende less so. And there occur larger, reddish balls of stilbite. They are strongly magnetic. Specific weight: 2.542.

Atkha [Atka] is the largest island of this group. Its shape is reminiscent of Unalaska and Umnak. Its SW end is narrow and low. Toward NE the island becomes wider and higher. Much like the Makushin Peninsula of Unalaska, the northern side of Atka, too, forms a peninsula occupied by tall mountains. The northernmost of them is a smoking volcano, called Korovinskoi [Korovin] Volcano. It is located near the sea at 52°24′ N. Lat. and 173°57′ W. Long., and has an elevation of 4852 ft. Four miles farther south, and a little farther east, rises Kliuchevskoi Volcano [Mount Kliuchef] (52°20′ N. Lat., 173°55′ W. Long.). There is a third, the Sarichef Volcano, not far from the east end.[79] And like the two others just mentioned, it is covered with perpetual snow. There are a number of other volcanos on the island, such as the Conical Volcano, 6 miles west of Korovinskaya Sopka (Konicheskaia Sopka, 52°22′30″ N. Lat., 174°06′ W. Long.). And again farther southward on a peninsula connected to the main island by a narrow land-bridge, there is Mount Sergeevsky [Mount Sergief] (Sergeevskaia Gora, 52°18′ N. Lat., 174°09′ W. Long.). The southern half of the island is mountainous, too. But nowhere here does the snow remain on the ground all summer long (cf. the view of Atka in Sauer, p. 182).

The Sarichef Volcano had a strong eruption in 1812. And in its surroundings there occurred such severe tremors that the people expected their imminent doom (cf. Vasil'ev in Golovnin's voyage, Vol. I, p. 173).

Korovinskoi Volcano has a depressed ridge when looked at from NW, and it ends in two peaks. On the north side it is very rugged and inaccessible, but not so from SW. The mountain itself is completely bare, and even several versts away there is no vegetation (1830). Everywhere the effects of earlier eruptions are evident, and the rocks are mostly black, porous, and easily crushable. Perpendicular rock walls arise in the surroundings of the volcano, and they consist of the same kind of rock.

Kliuchevskoi Volcano is after the last-named one the highest. It received its name, like the volcano on Kamchatka, because of the many hot springs which emerge from its west side. A few of them flow together and can well be used in a basin (for bathing). Ingenstrom finds the water similar to that in the springs of Sitka. Only the latter are more salty, taste more disagreeable, and smell stronger. After the water cools, it is covered by a thin, rust-colored layer, which allows us to assume a considerable content of iron. These springs are 5 miles from Korovinskaia Bay.

On the southern slope of the Conical Sopka, there, too, break forth hot springs with a temperature of 50° to 60° Reaumur. And farther up the mountain there are mud craters in the clay (Soupiraux). The Russians call them devil's ears (*chertovye ushi*). Most of the openings have an irregular, crater-like shape with a diameter of 1 *Tois.* to one foot at the mouth, and a hole the size of a fist at the bottom. Occasionally there are several craters together in one funnel. Several of these holes expel a hot clay of red and sometimes green, bluish, or light-yellow color; and this in intervals of one minute. Others are half or even brimful of it. And this clay is then constantly in a boiling, bubbling agitation in the manner (only stronger) of boiling pitch. Some of them are completely open, and only steam escapes them. But their interior does contain fluid, muddy clay (clay-porridge). Others, finally, are cold and dried out. In the surroundings of these mud craters there is a strong odor of sulfur; and a muffled subterranean noise can be heard, similar to that of several steam engines or iron furnaces. It is possible to walk on the grass or moss between these craters without any danger, although the ground quakes under foot and gives off a muffled noise. Only their slippery edges should be avoided. Those bare spots, which are covered with a kind of dry, fissured slake, are much more dangerous, because they collapse easily, and there is boiling hot clay underneath. Wherever a stick is poked into the ground, warm sulfuric steam escapes forcefully. A large number of these places are on the SW side of the smoking volcano.

Such a large amount of escaping heat, and perhaps also the mixing of the atmosphere with a certain amount of carbonic acid, cannot be without influence on the climate and the surrounding vegetation. A valley runs between Korovinskoi and Kliuchevskoi Volcanos and to the west of these mountains. It is watered by a river, and surrounded by high mountains. And it is not at all reminiscent of the rest of the sad, barren regions of the island. Fresh greenery, a profusion of flowers, and clear brooks which empty into a larger river, give this valley a more charming appearance than can be found anywhere else on the island. But one mile onward, this summer-like aspect becomes spring-like; and yet a little farther on we are surrounded by naked rock, snow, and on the whole the sad character of these regions.

Mr. Voznesenskii voyaged mainly about the northernmost part of Korovinskoi Bay. The following geologic conditions exist there, according to the specimens he sent:

From Cape Korovinskoi eastward, halfway toward Sergief Sopka, the coastal cliffs, rising to 300 ft., and a cave along this stretch, consist of an albite-porphyry, related to basalt, with cleavage like basalt, densely black, and interspersed with evenly distributed, small, and sometimes larger[80] crystals of albite. Outwardly this rock is reminiscent of the augite-porphyry from Rübeland in the Harz Mountains [Germany]. It distinctly differs from melaphyre, inasmuch as the black matrix (with traces of augite, olivine, and magnetic iron) contains albite instead of labrador. In front of the soldering tube this type of rock melts along the edges, to become heavy black glass, strongly magnetic, and with a specific weight of 2.736. Thus this porphyry is to be viewed as a link between rock-porphyry and melaphyre. The coasts of the northern Sand Bay [Martin Harbor][81] and of Korovin Bay consist of the same type of basalt-like rock. But this kind of rock appears as a greyish-green matrix with particles of augite and spiry [?] crystals, possibly of hornblende. Between this type of rock and its transitions, which resemble phonolite, there is found a loose, weathered, andesitic albite-porphyry. It consists mainly of albite[82] and of some grey matrix, which resembles that rocky porphyry. Near the top of Sergief Sopka, and on the northern slope of the same, there is a porous stone resembling lava, as well as real lava, and balls of lava, which have albite or sanidine and hornblende or augite in indeterminable proportions.

At the bottom of northern Sand Bay as well as on the slopes of Conical Sopka there are loose layers of rock and hardened deposits of clay containing fossils at an elevation of 30 ft. In the former there are probably *Cardium aleuticum* present, and *Nucula ermani*. According to Voznesenskii's suite, *Mya arenaria* var., and within the clay, *Cardium decoratum*, *Venus*, and *Tellina*, not well preserved. According to Dr. Blaschke, supposedly also *Ostrea plicata*. They belong to the most recent Tertiary age (cf. Appendix I).

The iron-containing hot springs of Conical Sopka seem to emerge from a vent in the clay-porphyry. The vent matrix or filler consists of a clay-like, water-solvent mass with a large content of partly deteriorated pyrite, breaking in parts (perhaps in the gouges) with quartz, as well as crystals of hornblende. The clay-porphyry has a grey, brown to red-brown matrix with more or less deteriorated crystals of feldspar and hollow bubbles partially filled with hornblende. On the surface this porphyry is quite weathered. Coal is supposed to be found near these springs, which, judging by a specimen, is baked into a red conglomerate of clay, wherein chunks of porphyry can also be observed. There are furthermore deposits of green jasper in larger amounts, and moreover reddish lithomarge, white argillaceous limestone marl, and well-deposits without a trace of infusoria.

Farther northward and higher up on Conical Sopka there are mud craters or devil's ears where Voznesenskii collected sulfur, red, light lava, black kings of feldspar-lava, slake, and obsidian.

On small **Solennoi Island** [Salt Island], at the southern entrance to Korovin Bay, 6 miles from Cape Iaichnoi [Egg Point], molybdenum breaks in gneiss, the same as on Unga Island. The Starichkof Rock-cliffs [Starichkof Reef] extend from Solennoi to the just-mentioned cape. In part they extend above the water.

Koniuzhii Island [Koniuji] is a mighty rock in a triangular shape. The surface is covered with pointed mountains, the shape of which changes often as the result of volcanic activities. From the midst of these rocks (52°13′ N. Lat., 174°54′ W. Long.) a thick smoke escapes from several locations. The Aleuts assure us that this land is rising, if slowly, but noticeably over the years. Thus there are locations once at water's level, where sea lions used to rest. They are now at mid-elevation of the rock (1827, Lütke).

Kassatochii Island [Kasatochi] (52°09′ N. Lat., 175°14′ W. Long.) is also only a mountain that rises above the water. On its summit there is a crater, about which the Aleuts report that it is filled with water. Seen from the north, the mountain top looks like a rounded back.

About the next group among the Andreanof Islands, the Chastiye Islands, we have very little information.

Great Sitkin Island is at 52°04′ N. Lat., 172°02′ W. Long. Its circumference is 25 miles; and from its middle rises a permanently snow-covered cone. Its elevation, according to Ingenstrom, is 5,032 ft. And according to Sarychev (II, p. 179 with profile) it belched fire at the end of May 1792. The shores of the island are rugged and strewn with boulders of rock (Lütke, *p.n.*, View 23).

Adak Island is large and mountainous, but not as high as Sitkin Island. The snow remains permanently in only a few locations. Sarychev, however, says (p. 5 and p. 188 Profile) that there are tall mountains, covered with snow. He did not notice any volcanic phenomena there (1770, the twenty-sixth of May). The island would probably be worthy of closer scrutiny. Shelikhov (p. 136) mentions Chekhina Island, about 40 versts east of Kanaga. It is reported to have a circumference of 80 versts.[83] But most likely he means Adak Island. On it, according to his report, are many mountains. Among them stands out the so-called White Volcano (Belaia Sopka;? 51°40′ N. Lat., 176°15′ W. Long., July 1784). In lower-lying locations hot springs break forth.

Kanaga
(Lütke, *p.n.*, View 24, and Sarychev, II, Profile to p. 80 and 188)

On its northern side, at 52° N. Lat. and 176°50′ W. Long., rises one of the tallest active volcanos of the Aleutian chain of islands. At the top it is depressed, and down to its middle it is covered with permanent snow. The south side of the island is flat. According to Lazarev and Andreian Tolstykh, the local inhabitants collect sulfur in the crater (summit?) of the volcano (in Schloezer's *Nachrichten von den neuentdekten Insuln zwischen Asien und Amerika* [News from the Newly Discovered

Islands Between Asia and America], Hamburg, 1776, p. 65; Busching's *Mag.* XVI, pp. 257 and 270). Shelikhov (p. 135), however, speaks only of the surroundings of the volcano, which he designated as fire-belching (Ognedyshushchaia Gora). According to him, there are hot springs at the base of this mountain, where the inhabitants boil their fish. According to Sauer (Engl., p. 226, German, p. 259) and Sarychev (II, p. 76) smoke was seen on the seventeenth of June 1791. It came from the hot springs of the formerly active volcano.[84]

Tanaga (Takawangha)

It consists of a high mountain ridge, which is divided into three groups: one on the east side, some distance from the coast; another in the middle, closer to the north coast; and a third at the SW end of the island, with one of the tallest volcanos of the Aleutian Islands. It falls off steeply toward the sea (Lütke, p. 322). The SE side is low. The highest points of the island are covered with permanent snow. According to Sauer (Engl. edit,. p. 221 and p. 181, with a beautiful view of the mountain; German, p. 251) the volcano, which he and Sarychev saw smoking on the ninth of June, 1791, is situated on the northwestern part of the island. Pallas (*N. B.* I, p. 297), Schloezer's *Nachrichten* (pp. 64–67), and Busching's *Mag.* (XVI, p. 270) report the same, provided that Tanaga and Takawangha are the same island. The circumference of the abruptly rising cone comprises almost ten geographic miles. That is almost as much as Aetna Volcano. The summit ends in several points, the highest of which smokes constantly. Permanent snow covers it downward past its middle. Therefore it is now always white. Its location is about 52°43' N. Lat., 178° W. Long. Ingenstrom (Lütke, loc. cit.) gives the length of the island as 25 nautical miles, i.e., 5 miles less than the sum of the earlier reports.

Gareloi Island (Burnt Island)

This westernmost of the Andreanof Islands consists of a mighty, smoking volcano, shaped like a pyramid (51°43' N. Lat., 178°45' W. Long.). Its summit ends in a sharp grade, which extends from N to S. This volcano rises so evenly from the sea that the coasts comprise its base. The island is formed like a triangle, the acute angle of which points south. The circumference amounts to about 18 miles. The mountain is covered with permanent ice (?) to over half of its height. Ingenstrom believes that this volcano, and those on Tanaga and Kanaga, are the tallest in the entire chain of Aleutian Islands, i.e., over 9,000 ft. high (Shishaldin). Sarychev passed the island twice (II, p. 65, p. 76, and profile to pp. 80 and 188), on the twenty-fourth of September 1790, and June 17, 1791. He mentions: "On it is a very tall, snow-covered, fire-belching mountain." (Sauer, p. 221, says the same.) [Merck, p. 62: no tall mountain mentioned. P. 161: dateline, June 18, 1791. A pair of tall mountains are mentioned here.]

The Rat Islands (Krys'i Ostrova; Khao)

They have the last known volcanos of the Aleutian chain of islands.

Semisoposhny [Semisopochnoi] or the Seven-Peak-Island, also called Unyak (Pallas, *N. B.* II, p. 321; Bragin), has a rounded shape with a diameter of about eleven miles. Its seven mountains, the highest of which is situated toward the ESE end, reach not over 3,000 ft. in elevation. Therefore, in summertime snow can only be seen in narrow stripes. One of the mountains on the northern side of the island (52° N. Lat., 180°15′ W. Long.) expels smoke (Lütke, *p n.*, Profile 25). According to Sauer (pp. 220 and 277) the mountain is pointed and lies in the southern part of the island. In his illustration there are several more smoking locations indicated, which, as Buch says, might have been cinder cones. In Sarychev (II, p. 179, with profile), it is merely mentioned that from a not very high mountain at the eastern end thick smoke is expelled (1 June 1792).

Amchitka

This largest and southernmost of the Rat Islands is low. It extends from NW to SE to E. Thus, its greatest width is the knee formed by these directions. According to Sauer and Sarychev (Billings's expedition, Engl., p. 151, German, p. 183; Sarychev, Russian, p. 4) the island begins in the west with a low spit of land, which gradually develops into medium-high mountains which comprise the main portion of the land. Toward N and W there are small, cut-off islands (24 March 1790).[85]

The cliffs of Kirilovskaia [Kirilof] Bay and part of the rocks of the island consist of rocks which have been changed by fire. They are held together by a dense clay. In several locations on the island there also is found brown coal (lignite), the wood-structure of which is preserved, according to samples sent by Voznesenskii. Voznesenskii furthermore collected: clay porphyry in a more or less deteriorated condition, kaolin, yellow ocher, and trachyte or andesite-like phonolith. This ferruginous, light-brown, dense rock contains very small crystals of albite, and needles of zeolite; and open spaces can be observed in it as the result of deterioration of some other ingredient. It has a specific weight of 2.576 and seems to be related to the trachydolerite found on Amlia and on Bering Island (cf. below). Voznesenskii furthermore sends pyrite, strongly weathered or again fairly well preserved, on veins of quartz in a talcose rock, and a coral-stock, undefinable in detail.

Twenty versts west of Amchitka lies **Sitignak Island**, according to Bragin (Pallas's *N. B.* II, p. 322). It is a small, rocky island with a volcanic mountain (1776) and a few hot springs.

On the most recent maps we find SW off the western point of Ajugadakh an islet without a name (51°43′ N. Lat., 181°38′ W. Long.). This is perhaps Sitignak Island, although it lies not far enough south, because Bragin states with regard to it: "It is unoccupied and without an approach; and 15 versts away from it lies Agadak Island (Ajugadakh, Krysii [Rat])."

Ajugadakh or Krysii Island [Rat Island] is mountainous. From its eastern end cliffs extend above and below water the distance of four miles into the sea (Lütke, *p.n.*, p. 326). Twelve miles north of there are **Little** or **West Sitkin** (Khvostov Islands) and Chugul [Sagula] Island. Both are mountainous; and toward the SSW end of the former, there is a cone-shaped mountain, the west slope of which has a smoking location (Lütke, *p.n.*, Profile 26). This volcano would be the last in the Aleutian chain, located at 52° N. Lat., 181°30′ W. Long. Formerly the volcano on Semisopochnoi Island was thought to be the last.

Kiska's southern half is of low elevation, and it is mountainous in the north. At its southwestern end there rises a rock in the shape of a perfect column (Sauer, English edition, profile, p. 218; Sarychev, II, p. 76; Lütke, *p.n.*, Profile 26).

Buldir, the last of the Rat Islands, is a uniform, tall rock, according to Sauer (German, p. 253; Engl., p. 219) and Sarychev (II, p. 76, profile). It is surrounded by many smaller rocks. Lütke says (*p.n.*, p. 327): "It consists of a low mountain chain, which sweeps three miles across, and five miles in length from NNW to SSE." It is located in 52°21′ N. Lat., 184°07′ W. Long. Voznesenskii received from this island edible earth or earth-cream, which contained no infusions. It consists of clean gypsum.

The Near Islands (O. Bliʒhniye)

They were first visited by Chirikov (Attu), later by most of the Siberian promyshlenniks. Lately the islands have been surveyed anew by Ingenstrom and Chernov. According to Teben'kov, the nature of the island's cliffs was described, but we do not have the description before us. Sauer (p. 218) and Sarychev (p. 76) have coastal profiles of Attu and Agattu.

West of Buldir are the Ingenstrom Cliffs [Ingenstrem Rocks], **Agattu** (Krugly), and the **Semischi** (Shemiyae) [Semichi] Islands. They are lone mountains that rise above the water level. The largest and westernmost of the Near Islands, **Attu** (Attak, Otma) is in part covered with cone-shaped mountains, the elevation of which does not exceed 3,000 ft. Volcanic activities were not observed. Chitchagof Bay [Chichagof Harbor], which cuts into the east end of the northern side of Attu, forms a beautiful harbor, the shore and bottom of which are sandy. In its proximity the Gavanka and Saraina Brooks (0.5 to 1 verst long, 7 ft. wide and 1.5 ft. deep), coming from small lakes, fall into the sea. The same is the case with Subienna Brook, which flows into Subienna Harbor [Massacre Bay] on the southern side. This is located not far from the eastern cape (cf. Zaikov in Pallas, *N. B.* III, p. 278). On its western point (Cape Wrangell) single rock cliffs stand alone (Sauer, p. 253), the same as on the eastern end of the island near Chichagof Harbor. Voznesenskii collected diorite, serpentine and aphanite, which resembles the melaphyre of Unalaska (north of the crater-shaped point). But it is lighter (spec. weight 2.231) and more nearly like diorite-porphyry.

The largest of the three Semichi Islands was called Little Alaid Island, probably because of its similarity with the Kuril island of the same name (Alaid or Herzfels, Serdtse Kamen, or Navel of Alaid, Alaidsakaia pupka, cf. Bering Island below). It is furnished with a bell-shaped mountain (Lütke, *p.n.*, Profile 28). According to Voznesenskii there are lapilli and volcanic sand here, with augite crystals and olivine.

CHAPTER SIX

➤➤ ◄◄

The Commander or Divided Islands [Komandorskie Islands]

ALTHOUGH THESE ISLANDS BELONG TO ASIA, according to their position, we nevertheless mention them here because of the similarity of their character with that of the Near Islands. Bering and Steller visited them first in 1741. Other visitors from the earliest times include Basov and Iakovlev, 1755; Krenitsyn and Levashev, 1768–1769, and Zaikov, 1773. In recent times Lütke gives us the most comprehensive account in his work.

Copper Island

Steller himself did not visit it, but he sighted it. According to him, the mountains there are lower than those on Bering Island. And on its NW and NE points there are many high and single-standing cliffs and pointed pillars in the sea (Pallas's *N. B.* II, p. 268).

Basov says in his report from his second voyage (1745–46, *Khronologicheskaia istoriia otkrytiia Aleutskikh ostrovov*, St. Petersburg, 1823, p. 4): "On the grassy shore (*laida*) we collected fifty pounds of good copper, and on the midnight side of the island two pounds of an unknown thing (*neznaemaia veshch'*), which is either ore or something else. And furthermore, my servants found here 205 large and small stones, among them two yellow ones, and one raspberry-red."

Petr Iakovlev, a foundry manger, who was sent in 1755[86] to seek the locations where copper had been found, reported as follows (cf. Pallas's *N. B.* II, pp. 203–308):

> On its northern side the island has mostly rocky shores, alternating with sizable bays. But on the southern side the shore is gentler and in part sandy. Only toward the southeastern point is the shoreline crowded with cliffs and sand banks, which at low tide are continuous with the shore. Eleven versts from the southern point, which is covered with small mountains and has a width of up to three versts in some points, the land becomes low in elevation and barely half a verst wide. Thus if the sea would rise a little higher, the point would become a separate island, while now it is connected to the larger island by that narrow

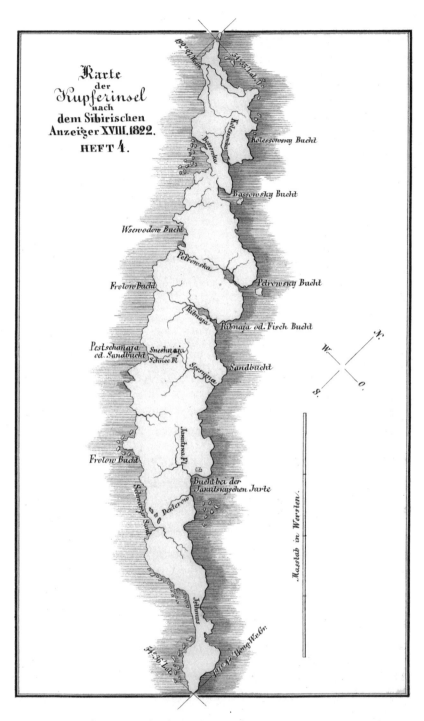

Map 4. Map of Copper Island based upon the *Siberischen Anzeiger*, vol. 18, part 4, 1822.

neck of land. No brooks fall to the sea from this southeastern section of the island. For a distance of 12 versts from the low neck of land, the island remains very narrow. And only in one single location, where at the north side there is a mountainous corner of land with cliffs off shore, it is up to 4.5 versts wide. But near the Yakutska Brook, which falls to the sea on the northern side 18.5 versts from the southeastern point, the land suddenly becomes 11.5 versts wide. Opposite the brook just mentioned, which flows from NW, as it takes up a tributary spring, there flow from the opposite bank three small spring creeks into that bay, called Frolova. From there on southward, the beach is covered with a black magnetic sand. To the north of there, there is a flat corner of land with seven or eight sand banks off shore. Thereafter the shoreline remains free and clean all the way to the northern point of the island. Another brook, Sneshnaia (Snow-brook), falls into a sandy bay on the same side, 5 versts from the latter. Opposite this one, in the north, a small brook, Sosnina, flows into a sandy bay, in a straight line about eight versts from Yakutska Brook. Right adjacent to this bay there follows a spit of land with a sand bank off shore, and then a deep bay, called Rybnaia (rich in fish). Into this bay empties a lake, formed close to the sea by Rybnaia Brook. The outflow is a short, wide channel. Into another, closely following bay, Petrovskaia, there flows a similar, even bigger lake, which receives the brook, Petrovska. This is the region, between the two bays and lakes, of the largest width of the land, 13.5 versts straight across. And on the southern side, toward Petrovskaia Bay, there is a good anchoring place, which is known by the name of Vsevidovskaia Gavan'. From this spring-fed harbor and Petrovskaia Bay to the northwestern point of the island the distance is estimated at 15 to 20 versts.

After the spit of land near Petrovskaia Bay, with the bank in front, there follows, 11.5 versts onward, a small but deep and navigable bay (Basovskaia Bukhta), which receives a spring-creek, Basovka, from inland. It was here the vessel anchored, which brought the miners. Close by is Kolesovskaia Bay, into which a small inland lake opens up, which in turn receives Kolesovka Brook at its upper end. Its location is only 10 versts from the point of the island.

From Petrovskaia Bay onward the land becomes narrower. Between the bays it narrows down to 5, 4, even 3.5 versts. And only at the spit of land that borders Kolesovskaia Bay, does the land still have some width. Then, because of the northward-extending points of land, the width is still four to five versts for a distance of five versts. Then it ends up with that point, the famous one known for its copper, which has given the island its name.

The entire island is bare of woods but full of mountains, which are very steep and consist of a brittle type of rock. Every year, therefore, when the snow melts, whole walls of it fall, especially along the coasts. All winter long great masses of snow hang above the steep rock ledges. And they fall off into the valleys—a great danger for hunters. Crosses can be seen in two places, at the Yakutska Brook where there is a cabin built in the Yakutian style, and near the Vsevidovian harbor. Their inscriptions testify that in the first location a Kamchadal from among Bakhov's people was killed by an avalanche on the seventh of April 1750. In the latter location a Kamchadal from Vsevidov's ship was killed when a rock-wall collapsed on the second of March 1747. And furthermore, while scraping at the copper-containing point, one mountaineer had his legs crushed by a falling boulder, which brought on his death a few days later.

The northwestern point of land, where the pure copper is found, runs out in a sharp promontory, which rises to a height of 25 to 30 *Klafter*.[87] Actually no copper ore, nor any pure copper on either of these slopes has been found. On the southern side of this reef, the shore-bank is 20 to 30 *Klafter* wide and flat; and at low tide it becomes exposed quite far out into the sea. In part it is covered with fallen rock-boulders, which are numerous especially on the northern side, too, where the shore drops off more steeply. There is no trace of copper here, either, except for the outermost point of the reef.

The outermost point of the island where the reef is barely 25 *Klafter* wide, exposes on the northern side on a steep incline, close to the baseline, two fractures, which are located barely twenty fathoms distant from each other. And the same distance separates them from the point of the reef. There are outcroppings here of narrow, northward-inclining fissures. The matrix is permeated by a mixture of green, shaly limestone with quartz and fractured spar. Almost all of the pure copper and copper-glass has already been hewn out of these fissures with mattocks. Nearby on the beach, exposed by low tide, small pieces of copper, the size of beans, which the sea has ground smooth, were collected. On the southern side of the point of the reef three fissures were found in different locations to a distance of about 100 *lachter*[88] from the point, on the low shoreline, partially below the line of high tide. Fifty pounds of pure copper was mined at that time and place in the shape of various pieces, leaves, and masses. And there is a fourth location on this side, 150 *lachter* from the point of land, right at the sea, where small fissures of copper and copper-glass crop out in an area 7 *lachter* long and 1 *lachter* wide.

Pallas (l.c.) remarked:

The report before me states nothing definite about the type of mountain common on the island, nor of the type where the copper-bearing fissures occur. Among the specimens that have been brought from there, is one which is grey, clay-like, more or less calcareous, which, however, does not react. It is permeated with little fissures of spar. The largest piece of pure copper known to me, which comes from there, is in the cabinet for natural specimens in the Academy of St. Petersburg. It weighs ten pounds and has the form of a shapeless mass, almost melted and partly ground smooth by the sea. Thus, most of the pieces obtained from Copper Island are sometimes the size of an egg, but most of them no bigger than beans or nuts. Some of it sits in and on this kind of rock in different-shaped leafings. But I have two quite well preserved small kidneys from this Copper Island. The inside shows a dendritically developed copper with densely mingled points. Among the small pieces can be found quite an amount of coarse, red copper-glass, some with, and some without pure copper, with or without calcspar fissures. On the whole the pure copper has reportedly become rare on the island, where at 3 to 5 rubles per pound, people in Kamchatka formerly fashioned it into various small ornaments, bracelets, and the like.

Krenitsyn's report (Pallas, *N. B.* I, p. 253) is as follows:

On the NE coast, the only one usually visited, much pure copper has been found. It is washed up by the sea and lies about on the beach in such amounts, that ships could be loaded with it.[89] The island is not high, but it has many pointed mountains, which have the aspect of dormant volcanos. On the whole it should be remembered at this point, that all of the eastern islands represented on Krenitsyn's map, even the smallest of them, show such pointed mountains (Sopki— pointed mountaintops?!). Some of them consist of one single such cone, which rises from the sea. Thus the entire chain of islands can be viewed as consisting of such volcanos, without the need to do violence to one's imagination. Everything there seems to indicate a newly evolving land. All these islands are subject to strong earthquakes; and on all of them sulfur can be found in large amounts. The journalist among the Russian argonauts was unable to say whether lava was not also found on them. He speaks of a kind of colorful rock as heavy as iron.[90]

Zaikov (Pallas, *N. B.* III, p. 277) gives us no new information either. "On all sides of the island," he says, "there are steep rocks with an elevation of approximately 40 to 50 fathoms. And on the northern side there are small bays and two small rivers. On the western side of the island the sea, when it goes high, is washing up pieces of copper; and the side of the promontory looks like a copper mine. Snow lies on the mountains until July."

Lütke describes the island as follows (*p.n.*, p. 334):

Among all of the islands of this ocean Copper Island is unique because of its narrow shape. Only Amlia has a resemblance with it. Its length, incidentally, is 3 nautical miles (7.5 German miles or 50 versts). Nowhere is it over five miles wide; and it gives the impression of a mountain range, which sweeps from SE to NW, the summits of which extend above water level. Seen from Bering's Cross (on Bering Island), it appears as if there are three islands. Its elevation is little less than that of Bering Island. According to Golovnin, (I, Chapt. V; and II, Chapt. III), its SE end lies at 54°32'24" N. Lat., 168°09' E. Long. The NW end is at 54°52'24" N. Lat., 167°31' E. Long.; while according to the map of the Hydrographic Institute at St. Petersburg, 1848, the SE end is indicated in 54°36' N. Lat., 168°12' E. Long., and for the NW end 54°55' N. Lat., 167°33' E. Long. There are no volcanos on Copper Island, but earthquakes are frequent, and often of long duration. Thus in June of 1827 the swaying of the ground was felt for four minutes without interruption. During an earthquake, too, the ocean often rises suddenly to ten feet high, and just as suddenly recedes.

According to Erman (Berghaus, *Annal.*, Vol. VI, 1832, p. 455) there are out-croppings of genuine mining stone (Zechstein) on Copper Island. And in his travelogue he says on March 1, 1848, p. 525, that the copper of this island is a char-acteristic ingredient of the graywacke formations, which are transected by the andesite summits. "I have received," so it states in l.c., p. 559, "from Copper Island pieces of quartz with green oxydized ores, as well as a tallow-containing, coarse limestone, which is permeated with copper-green and malachite, much like a similar type of rock in the Ural Mountains near Nizhne Tagilsk (January 1, p. 349)."[91]

Voznesenskii's specimens show trachydolerite as the predominant rock from the NW end of the island (Kolesovskii Bay). The same is also found on Amchitka Island and on Amlia, as well as Bering Island. Its description is found with the latter. Pure copper and copper-glass break in this type of rock or in metamorphic shales on ducts of calcspar and quartz; copper pyrite in red jasper which resem-bles pitch stone.

Bering Island

Bering Island extends, according to Lütke (*p.n.*, p. 331), from SE to NW between 54°41'05" and 55°22' N. Lat., and 166°43' and 165°[20'] E. Long. Its length is 5 nautical miles (12.5 German miles or 85 versts); and in the widest place it mea-sures 16 to 17 miles. According to Steller's reports, the island must once have been almost twice as large, because it is hardly believable that the translation of 23.5 Dutch miles into 165 versts (Pallas, *N. B.* I, pp. 255–301) indicates an error.[92] The exacting Steller could only have meant the German mile, which would amount to 22 versts, not the Dutch, which at 100 *Roden* equals 0.937 verst. We shall take up Steller's description, so far as it belongs here, inasmuch as all the later reports (Krenitsyn in Coxe, p. 252; Kotzebue, I; Golovnin, I; Beechey, I, p. 377, German) derive from this source.

The island itself (1741–1742) is 23.5 Dutch miles or 165 versts long, and of differing width. The southeastern point is for the distance of two miles in a western direction only three to four versts wide, up to the place we called Neobkhodimyi Utes (The Unavoidable Rock); from there to Sivuchia Guba (Sea Lion Bay) five versts; at the Bobrovyi Utes (Sea-beaver Rock) six versts; from our dwelling, directly across, seven versts; near Kitova Reka (Whale Rivulet), where there is a large bay in the south, again five versts. Near Lesnaia Reka (Bush River) eight versts. From there on, the width of the island increases very gradually, until finally the largest width of the land is 23 versts or 3.5 miles toward Severnoi Nos, or the promontory extending northward, which is 115 versts distant from the south-eastern point. From here onward the land again turns toward NW and gradually diminishes in width. Thus 135 versts from the southeastern point it measures only 5 versts; and 15 versts onward only 3 versts across. And so it continues until it diminishes at the other point down to a minimal width of one verst. Thus the width of this island has a very unequal relationship to its length. The same is the case with all the other islands we saw about America and in the channel.

Bering Island is a row of barren cliffs and mountains, connected with each other. They are divided by many valleys, which extend south to north, but rise as a single rock from the sea. The highest mountains are in vertical elevation not over 1,000 fathoms high.[93] They are covered 0.5 *Schuh* (foot) deep with a common, yellowish clay, below which there is found a layer of poor, yellowish, deteriorated rockstone, two to three feet thick. Then begins the uniform rock, which reaches into the depth that can be observed at the steep cliffs of the shoreline. In those parts which are exposed to the sea in the north and south, the mountains are dense and undivided in consistency. But those open to the interior because of the east-west valleys, are cleft apart and weathered because of the moisture which breaks them up with frost in wintertime. The mountains are all equally sweeping from NE to SW. The valleys with their brooks and springs all open northward and southward to the sea. And the latter flow from their source in the SE toward the southeastern end; from NW, however, to the northwestern end. This I have observed throughout the duration of my travels about that land, as I have from day to day recorded it in my journal. Even and level places are nowhere to be found in the interior, only high mountains and narrow valleys. But since the valleys are mostly filled with brooks, one is obliged to seek his way across the lowest mountain ranges, if a crossing is to be made from one side of the island to the other. This was the more difficult for us and toilsome, as laden with animals and fish, we had to pick our way across. The only level places can be found along the shores of the sea, where in a half-circle the mountains recede half a verst, or at most one verst from the beach. And such locations often occur where brooks are found. That is to say, a constant relationship can be observed, that wherever a mountain range runs out, northward or southward, to form a spit or angle of land, the shoreline behind the same becomes even and wide. The steeper the corner of the mountains, the smaller is the plain behind it. The more gently, however, the mountains fall off, the larger is the plain behind them. This is also the case wherever the mountains sweep along the land from SE to NW. The larger the plain, the lower the mountains, the stronger are the brooks that flow from there. The more steeply, however, the mountains approach the shoreline, the smaller, if more numerous, are the brooks. Where shore and mountains fall off abruptly toward the interior, if they are firm and cohesive, their likes can always be found one or one-half verst from shore. They empty into the sea through brooks. The cause for this seems to be that water, created by snow, rain and fog, comes down with such a force that it washes out the soft upper layer down to the bedrock, which constitutes the basin of these lakes. Thus, because the mountain range retreats in such locations, the springs emerge at the foot of them and provide room to form such lakes. Their origin can thus be explained, which is very different from that of lakes in large flatlands. They usually have a muddy, gluey bottom, but wherever the range descends gradually, there the waters form a continuous valley, and at the bottom of it a brook.

All the mountain ranges of the island consist of a common, grey rock (granite). But wherever they run parallel to the sea, the point where the mountains extend into the sea usually changes and becomes a clear, greyish, firm sandstone, good for use as a grindstone. This is a phenomenon which seems very noteworthy to me, inasmuch as it appears as if a transformation of the rock, the structure of which is very unusual, is the result of its exposure to the sea water.

In many places the shore beneath the rocks is so narrow, that at high tide one can just barely get through. A few places can only be passed during low tide. But in two locations it is patently impossible. One of these is not far from the northwestern point of the land. These might originate from earthquakes, large tidal waves, wash-out of the ground by the waves, and cleavage of the mountains by way of frozen water. Evidence of this is in part apparent in large heaps of rocks, and partly in the pillars and rocks, torn from the shores, which stand alone in the sea. These can frequently be found in such places. The southern side of this island is on the whole more rugged, more stony and full of cliffs than the northern side. On the northern side it is possible to walk everywhere along the shore, except near Neobkhodimyi Utes and behind the spit of land extending northward, which is very steep and at the shore full of cliffs and fallen boulders of rock. Here and at other places I have found strange sights and plays of nature beneath these rocky ruins. Thus, for instance, near the Peshchera Cave (Steller Cave), which I named, the mountain resembles quite naturally a wall, and the ledges bastions and other fortifications. Behind the cave is a larger number of single cliffs here and there along the shore. Among them are some that represent pillars, vaults, and arches; and it is possible to walk beneath some of them.

I have also noticed this difference between the opposite shorelines: wherever there is a bay on the northern side, such as the one near our dwellings, on the opposite, southern side a spit of land extends out into the sea. If the region of the northern shore is wide and sandy, then the opposite shore on the southern side is that much narrower, rockier and rugged. If, however, the northern shore is such that it is hardly possible to pass, then the southern side is found to be that much wider, level and sandy.

Caves and ravines, the result, it seems, of earthquakes at different times, are variously encountered in different places. That cave just mentioned, which is named after me, and Iushin's Shcherlopa are the largest among them.

On the highest ranges and the uppermost summits of them, I have noticed that from their middle there extends a kind of heart or core, which ends in a bare, conical, upright rock. It does not differ from the other type of rock, but it is smoother and cleaner, and it has the shape of a determined figure. I have encountered such points of quartz in 1739 in the mountains at Lake Baikal and on Ol'khon Island within that lake. Another type of rock, green, almost like malachite, somewhat translucent, and fibrous like stalactites, was sent to me with a message from Anadyrsk. There, too, they crop out at the summits of the mountains; and it is reported that they even regenerate wherever they break off. If this were so, it should be possible to explain it as the result of pressure from the inside.[94]

When the land suddenly changes directions and abruptly moves to a different region, I have always noticed that the shores are very rocky, one or two versts prior to that location. The mountain ranges sweep outward to the shores. They become very steep; and at their outer points they are split into single, broken cliffs and pillars. Bourguet, incidentally, has observed in the Pyrenees Mountains that the surface of the mountain ranges has certain regions. Thus, they possibly reveal their origin from the sea. I, too, have observed the same, not only on the mountains of this island, but also throughout Kamchatka and

Siberia. Consequently, whatever he has noticed concerning the formations of valleys and of the spit of land opposite the bays,[95] I find applicable to this continent as well, as the result of changes that occurred gradually through flooding, earthquakes, and other chance occurrences.

Concerning the shores of the ocean around this island, they are very peculiarly composed. Thus it can be said without arousing any suspicion, that we have been preserved in that land and saved from utter destruction by a miracle of God. Although the island is 23 Dutch miles long, not a single location can be found along the entire northern side, which is suitable to harbor even a small vessel. The shoreline is littered with rough cliffs and rocks, two, three, and in some places up to five versts out into the sea. After the water runs off at low tide, it is possible to walk out, dry-shod, for many versts, which are all then covered with water again. And at falling tide the waves rush at the cliffs with such a roar that we could often not behold the sight from land without a sense of horror. The sea becomes so full of foam from the frequent impacts, that it looks like milk. Only one single, narrow passage has become known to us on this side, which is clear enough of cliffs that it is possible to anchor there when the sea is quiet. This was the very space, about 80 fathoms wide, to which God led us providentially and lovingly, when we blindly, wearily, and desperately ran toward shore and destruction under full sail. And through this portal we were led out again as well. At this very place is the largest cove along the entire north side.

All these circumstances make it plain that in former times this island was much larger and wider than it presently is. And the limit of the former size of the island can be perceived from the many cliffs and ruins which lie out to sea. Three possible grounds for proof throw light on this observation. Firstly, the rock cliffs in the sea maintain the direction of the mountain ranges on shore. Secondly, the brooks that run off though the valley maintain a clear channel also in the sea. Thirdly, the veins and deposits, which show up blackish, greenish, or white like quartz, on the cliffs in the sea, can be followed uninterruptedly on land to the base of the mountains—a sure sign that they had once been a unit with the land. Fourthly, it usually is a rule that where the mountain range descends gently toward the shore, or where the land is on the whole low and the coast sandy, there the sea is also shallow; and only by and by does it gain in depth. But where the shoreline is steep and abrupt, the sea is at once quite deep close to shore; and at a distance of twenty fathoms it often gains 60 to 80 fathoms of depth. But beneath steep cliffs of this island the sea is no deeper than in other places, because the bottom is filled up with cliffs that have collapsed. Finally, we ourselves have been witnesses of the gradual diminishing of this island. In the winter of 1741, for instance, a considerable portion of the mountains was washed out near Iushini Pad' (Iushin's Valley). In spring, burst open by the frost, it collapsed by itself. And it so happened that on the eighteenth of June I had walked westward beneath a cliff by the sea. But when a few days later I returned, I found that during this short time the entire rock wall had collapsed into the sea; and thus the entire region had gained a different aspect.

The seacoast of the southern side of the island has a much rockier and more rugged aspect; there are nevertheless here two locations where it is completely safe to go beneath [*sic*] the land. And with small vessels, *Scherr*-boats for instance, it is possible to navigate into the estuaries of the rivers, or rather lakes, which empty into the sea by a short channel. There it is possible to stay as in a harbor. The first such place is located seven Dutch miles from the southeastern spit of land, within a large bay. It can be located from far out at sea by the unique stone pillars on its western corner. It is this very place we have called Iushini Pad' (Iushin's Valley) after the first discoverer of the same, Helmsman Iushin.

The other place is located 115 versts from the southeastern spit of land and 50 from the northwestern. It is even more recognizable, because in just that location the land turns from north toward west. And in that angle a little river opens out, which is the largest of all on the island. At high water it is six to eight feet deep at the mouth. This river comes from the largest lake on the island. From the ocean toward the lake it becomes progressively deeper. Thus, it is possible to proceed without difficulty through the same the distance of 1.5 versts from the estuary into the lake. There it is the safer to stay, as it is surrounded by rocks, like walls, which are a shelter against the wind. I have called this river Ozernaia; and it is the more conspicuous as there is an island in the south, opposite the estuary. It is one mile in circumference, and it is only one mile distant from the river mouth. The westward shore is sandy from there on for five miles, level, and flat; and the sea is clear of cliffs. I have never been able to detect in either rising or receding tide, such movement of the waves (*burun*, surf) as usually reveals hidden cliffs, although I have remained there with that intention for three days.

From the highest mountains of Bering Island, two islands can be seen toward the southern side on a bright and clear day.[96] The size of one of them is about one mile in circumference. Its shape is oblong. Its location is 50 versts or 7 miles from the northwestern point of land on Bering Island, and 1 mile from the southern point. The other one consists of two high, fractured cliffs in the sea. They have a circumference of two or three versts, and the distance from the island is about two miles. This latter island is located straight opposite the northwestern spit of land, due SW.

From the northwestern spit of land itself, very high, snow-covered mountains can be seen in the NE, at a distance of about 15 to 20 miles. I take these for a promontory of the American mainland itself, rather than an island. Near our quarters on the northern side, too, such high and white mountain ranges have many times been observed at the same distance. A few times it was even possible to discern another island southeastward; however, very indistinctly. Even during the clearest weather I always observed a haze west and southwestward over the land of Kamchatka; and thereby I have deduced the proximity of that land.

[Addition from the errata sheet for the original edition, p. 423:] Addition to Bering Island from Steller in Pallas, *N. B.* II, pp. 270 and 271:

Earthquakes occurred three times. One of them occurred in conjunction with a wind storm on February 7 at 1:00 p.m. and lasted all of six minutes. I was in our quarters below the earth's surface. There I perceived, among other sensations a few minutes prior to its onset, a noise of a strong subterranean wind, which seemed to proceed from south to north with a strong hissing and rushing. And it became ever stronger the closer it came. After the rushing ceased, the quaking started, which was so strong and noticeable that the pillars of our quarters moved and everything began to crack. I immediately ran from the house down to the sea, in order to observe what was happening in nature. Although the shaking on land continued unabated, I could not notice the minutest unusual motion in the sea. The air, by the way, was bright and clear, and the weather pleasant. Another earthquake occurred on the first of July toward evening at five o'clock. The weather was very clear and pleasant, with the wind standing easterly. The largest alterations on this island seem to be the result of earthquakes and high water-floods. There are clear indications of mighty deluges, such as driftwood, whale-bones and whole skeletons of sea-cows washed far inland between the mountains. From the age of the wood, I have been able to deduce quite clearly that during the deluge, which in 1738 affected the coast of Kamchatka and of the Kuril Islands, on Bering Island, too, the water had stood 30 fathoms high. Not only whole trees testified to this fact, which I had observed on the mountains in such vertical elevation, but also the sand hills and new mountains washed up not far from the beach. Large trees protruded from them, as yet undeteriorated. While observing these hills, which originated from flooding, I found it very noteworthy that the shape, location and number of summits and valleys corresponded exactly to those of the high hills, at the bases of which these had recently arisen. Thus, it is most likely that the higher mountains owe their formation and cleavages like-wise to the power of the waves. [End of addition from p. 423]

Erman deduces from descriptions and rock specimens which he received (Berghaus, *Annal.*, Vol. VI, 1832, p. 455), that there are graywacke and terenite formations on Bering Island, which seem entirely analogous to those in the Als-danian [Aldan?] high mountains. But according to Voznesenskii's shipment there is an outcropping of trachydolerite on the west coast of the island, that of Copper Island, Amchitka, and Amlia. This rock is finely crystalline, tough, not hard (3). It feels rough, looks grey to greenish; and its composition is predominantly white balls of zeolite, light-green hornblende, and magnetic iron. Mica occurs rarely, and there are some traces of augite and albite; but their presence cannot be positively confirmed. All these ingredients form a dense mixture, and small points of zeolite give it a weathered appearance. Before the soldering tube this obviously volcanic type of rock, which does not have the characteristics of lava, melts with difficulty along its edges to form a pearlstone-like glass [perlite]. It is strongly magnetic, and the zeolitic components dissolve in hydrochloric acid. The specific weight is 2.712. The large weight is probably caused by the magnetite, in spite of the predominance of the zeolite. The andesite has the same specific weight. But nowhere was in this case a predominance of zeolite indicated, because on the whole rocks of that type are not considered to be andesite. Should hereafter the occurrence of andesite on the island be proven beyond a doubt, and this name consequently be adopted as characteristic, then these rocks will have to be classified as andesite-dolerites, or else as andesite-like phonolite (analogous to the trachyte-like phonolites). The assumption is that he mistook andesite for granite. Our rock type could easily be the "greyish, firm sandstone" at the coast, which Steller presumed to have been transformed by contact with the sea water from that grey kind of rock. Iushin's Shcherlopa, Steller's Cave, and the rocks beyond it, composed of pillar-like vaults and ruins of walls, make it probable that at least in this location volcanic or eruptive kinds of rock are present. Sandstone seldom occurs in such forms. They are not known in either Kamchatka or the Aleutian Islands. Noteworthy are the sand hills, mixed to a height of 30 *Klafter* above sea level with driftwood and skeletons of sea animals. At the present time they are assumed to be the result of flooding, caused by volcanic eruptions in the sea off Kamchatka and Japan. According to Steller's observation of wave formations, they could just as well be sand dunes raised in the process of the rising of the island. The cones that protrude from the highest mountaintops like hearts or cores, which resemble Little Alaid Island from among the Near Islands, as well as the Kurilian island-volcano Alaid or Heart Rock (Serdtse Kamen), or Navel of Alaid (Alaidskaia pupka: Erman's *Reise*, I, 3, p. 525), lead us to presume the presence on Bering Island of the same elevation craters with the same cones of trachyte and andesite rising up from them. The same lodes of white quartz, which Steller mentions, can also occur in clay-slate, which might become apparent as the results of the collapse of the embankment. Now that we have looked at all the islands located in a half circle between Asia and America, we shall now turn to the Pribilof Islands, which are a part of the administrative district of Unalaska.

CHAPTER SEVEN

->- -<-

The Pribilof Islands (Ostrova Pribylova)

THE PRIBILOF ISLANDS are located on the longitude of the western Fox Islands, and on the latitude of Mt. Chiginagak on Alaska, and of Edgecumbe [Kruzof] Island. They were discovered by the helmsman Pribylov in the first days of June 1786 (St. George), and on the twenty-ninth of June 1787 (St. Paul). At first they were called the New Islands; then the Lebedevskie Islands. Thereafter, She-likhov called them Zubov Islands (after Prince Platon Aleksandrovich Zubov). Later they were called Fur Seal (Kotovye) and the Northern Islands (Severnye). Finally, Sarychev gave them their present name, while the colonists usually call them the Little Isles (Ostrovki). The group consists of two large and two small islands. Before the Russians discovered them, they were uninhabited and unknown to the neighboring peoples.[97]

St. George (Lütke, *p.n.*, Views 30 and 31) has a length of 13.5 miles and a width of 3 miles, and it extends fairly straight from west to east. From the southeast side the island presents a uniform aspect, appearing as a plateau of medium elevation with one somewhat higher point, about 1,083 ft., Engl. The outermost ends of the island consist of very rugged rocks. Those of the north coast are 300 ft. high (150 ft. according to Postels). And they rise straight out of the sea. Only in one location does the coast have hilly slopes, 5 miles from the NE end (56°38' N. Lat., 169°10' W. Long.), where the settlement of the [Russian] American company is located. The few small bays do not cut deeply into the land.

Above the naked rock walls, horizontal strata of lava can clearly be observed. They are probably not clay-porphyry, as Kotzebue indicates in his first voyage, Part III, p. 168. They are traversed by many straight and oblique fissures. Everywhere in high and low places boulders or porous lava are lying about. Postels received from an Aleut a dense basalt with grains of olivine. Lava and slake show that once there was an eruption there. Fires have also been observed several times from the islands toward NE across the sea (Kotzebue, l.c.), and the inhabitants are of the opinion that an eruption cone is forming there.

Khlebnikov says that the island consists of granite and gneiss; but it seems, as it often happens, that he mistook other rocks for these. Beechey simply states (twenty-first of October 1826): "St. George's consists of two hills, united by moderately high ground, and is higher than St. Paul's." Veniaminov says (I, p. 273): "On St. George there is little clear evidence of volcanic activity. Everywhere (p. 288) there are deposits of granite, gneiss, and hardened clay. And on the mountain heights there is mica, and on shore garnets and pyrite." But in the same place the following is stated (p. 291): "The earthquake on St. Paul (second of April, 1836), was more violent on St. George where the rocks burst apart and collapsed." Voznesenskii, who touched three times on that group of islands, collected the following types of rock:

Much red and black lava, very porous, and the pores not filled. The red lava is found on the northern part of the island. At 1,000 ft. of elevation, there are a large number of boulders of black lava lying about, which look like slake. The lava streams at the ocean level have above the level of the water a thickness of 6 ft. The same lava is also widely distributed near the settlement. It seems that the latter rests upon the lava. On the northern side of the harbor there is an outcropping of basalt with olivine (basaltic lava). And in trough-like deposits between them there are the largest masses and nests of olivine, which Voznesenskii has sent in large amounts. The outer surface of the same is weathered and decayed in a well-known manner. Within these pockets of olivine there is also hornblende (according to the measurement of the angle of 124°30′), which is noteworthy, because it supports the transformation of augite into hornblende (uralite). Other pieces indicate hornblende and glassy feldspar in soft, but very tough, grey claystone.[98] A blue potter's clay, with pieces of talc enclosed, is being brought to the island of St. Paul, because there it does not occur. Boulders: bluish, phonolite-like rock with small crystals of spinel, or also hyacinth, analogous with deposits on St. Paul.

St. Paul

(Lütke, *p.n.*, View 32; Sarychev, I, p. 86)

Langsdorff, who was on the Pribilof Islands from the second to the eighth of July, old standard, 1805, describes St. Paul as follows (Vol. II, pp. 17, 21): "The island consists of several low mountains and hills, in the valleys of which many freshwater lakes can be found, the result of melted snow. They yield good drinking water. The island is 30 to 40 versts long. (According to Veniaminov 25 versts long, and 18 versts wide.) And it extends from NE to SW. It is overgrown with grass and low shrubs, but completely bare of trees and bushes. The NE end runs out into a low cape, which leads into a wide-open, shallow bay. The sand along the shallow coast was black, shiny, heavy, probably with contents of iron or titanium. It seems to me that it consists of crushed lava. Wherever the shoreline rises higher, the walls are composed of horizontally running layers of lava. The steep cliffs of the southwestern coast consist of a brittle, porous, black lava, which at once leads to the thought that this island owes its genesis to a volcano. I received several fossils from the highest mountain there, which is located approximately in the middle of the island."

According to Lütke the eastern and northern sides of the island rise gently, while the west side is mountainous and ends with a high, precipitous cape. On the east side there is a mountain—not very high—the summit of which seems to be caved in. Beechey also mentions (II, p. 59, October, 1826) that one of the three small peaks on the island has the appearance of a crater. According to Veniaminov (I, p. 280) a range of hills traverses almost the middle of the island from W to E. The eastern end of it forms the highest point. And on the top of that mountain there is a small lake, filling the above-mentioned crater, noted by Beechey and Lütke. And furthermore, on the western end of that range of hills (Veniaminov, l.c.), conspicuous crater formations are found with evidence of activity which was only recently interrupted, namely: burned rocks that were baked together. Veniaminov believes that St. Paul must once have consisted of several individual islands. As proof he points out that recently a spit of land near the settlement (57°05′ N. Lat., 169°51′ W. Long.) was in a short space of time dried out. He attributes this phenomenon to the amount of sand thrown out by the sea. A slow rising of the land, on the other hand, is not improbable. According to Chamisso (Kotzebue, I, part 3, p. 168) the peninsula where the settlement is located consists in part of volcanic cinders, and in part of a lava resembling iron slake, the coarse surface of which is still without vegetation in several places, therefore there is no doubt that once it was under water. In the vicinity of this settlement on the SW side, Voznesenskii collected black basalt with sparse feldspar needles and clear primary cleavage from a rock wall, 30 fathoms high, and from beneath the rock masses that are grouped like foundations beneath a house. There is also red and black lava and lava bombs with a more or less dense matrix, fine feldspar needles and olivine; and then olivine with hornblende, hyacinth and feldspar in pockets like those on St. George; rock crystals and pyrite. On a mountain on the east side of the bay, finally, where the village is, Voznesenskii found an undefinable fragment of coral, and besides that a dark-grey lime marl, which also contained clay-slate; and in a lime-wacke or conglomerate with lime-like adhesive, the following Tertiary fossils: *Cardium decoratum, C. groenlandicum, Venerupis petitii* var., *Pectunculus kaschewarowi, Astarte corrugata, Tellina lutea* (cf. Append. I).

Mammoth teeth were found here or on St. George in the year 1836 (Veniaminov, I, p. 106). And on the second of April of the same year a strong roar was heard on St. Paul Island; such a strong earthquake was felt that people could not stay on their feet (cf. Khlebnikov's entry in Baer and Helmersen, *Beiträge* I, p. 325; Chichinev on the earthquake on St. Paul Island, in the same place, p. 315; and Veniaminov, I, p. 291). Mountain rocks burst and crumbled. The subterranean roar progressed in a direction from E to W. In August, too, a similar din was heard, but much diminished and muffled.

On both sides of the SE end of St. Paul there are three small rock islands: Beaver [Otter], Sea Lion, and Walrus Islands. The first has a circumference of six versts. It is high and rocky, and low only on its northern end. Everywhere traces of volcanicity can be seen; and the water of a small lake on the island, which occasionally dries out, has the taste of saltpeter whenever the surf is strong (Veniaminov, I, p. 285).

Regarding St. Matthew, St. Lawrence, and St. Diomede Islands, see their earlier treatment, which is located with the coastal description of the mainland.

CHAPTER EIGHT

Volcanic Phenomena

NOW THAT WE HAVE ACCOMPLISHED the orographic and geognostic description of the west coast of North America and of the islands between Asia and America, so far as the scanty material allows, many a reader might be led to remark that the reason for the work before us was primarily the volcanic character of these regions, except for the specimens sent by Voznesenskii. It is, indeed, only the closer investigation of volcanic phenomena which has given geologists a deeper insight into the composition of the earth's interior in the first place. Only on the basis of that research, which provides us with enough subject matter whereby to compare other, less-known phenomena acting according to the same or minimally modified laws, can the geologic material we have, incomplete as it is, be made useful. For the study of volcanism it is necessary not only to investigate the rock conditions and the composition of the mountain ranges, but also to have an exact knowledge of the behavior of volcanos, the changes within them, their periods of activity or rest, etc. Historical evidence is therefore not without importance. The latter was not much acknowledged in the works of Hoff, Buch, Hofmann, Lütke, and Veniaminov, concerning these regions. And in other works, which treat geology only incidentally or to make an occasional point, historical evidence is absent altogether.[99] But since we have access to most of the sources for the history of the Russian American colonies, a corresponding acknowledgement of the same was very much in order. From this source also derived the chronological overview of the voyages given in Appendix II.

There is hardly a more grandiose theater of volcanic activity than that of the Aleutian Islands, Alaska [Peninsula], and the western rim of Cook Inlet. And our century gives us here an insight into the entire complex of phenomena, from the most grandiose expulsions of force, such as the rising of mountain ranges, the sinking of large surfaces, earthquakes, eruptions of lava, expulsions of cinders, ashes, and mud, to the weakest: hot springs and expulsions of gases, or exhalations and sublimations of various kinds. The geologic character of most of the islands here observed, and of a part of America's mainland as well, speaks predominantly of volcanic elevation and origin; and earthquakes have been felt here almost everywhere. Nevertheless, from twenty-five Aleutian

Islands we have furthermore received information about other volcanic phenomena as well.

In the group of the Rat Islands: three islands
In the group of the Andreanof Islands: nine islands
In the group of the Fox Islands: thirteen islands[100]

These islands have forty-eight locations of volcanic activity, not counting the hot springs and mud craters. Past volcanic activity is probable on the Pribilof Islands, on St. Lawrence, and on St. Michael. Furthermore there are on:

Alaska Peninsula: four volcanos
Cook's Inlet: two volcanos
Chugach Bay [Prince William Sound]: one volcano
Copper River: one volcano
Kruzof Island: one volcano

Three other supposed volcanos, located between the Copper River and Mt. Edgecumbe, and more of the same on Prince of Wales and Beaver Islands, are not exactly known as to their number and location. They are doubtful, and at any rate, have not been active since the discovery of those regions.

The distance from the Wrangell Volcano to the volcano on Little Sitkin Island, between 62° and 52° N. Lat. and 143° and 181° W. Long., is 376 geographic miles or 1,505 nautical miles.[101] Among all the volcanos along that distance, the Wrangell Volcano is the only one "standing isolated, and not connected with a mountain range," and is therefore to be viewed as a central volcano. For the distance, however, from that mountain to Mt. Iliamna (or High Mountain), and from Iliamna to the Pavlovskaia Sopka (95 and 97.5 geographic miles), the region is very little known.[102] Thus, we cannot positively determine whether there is not after all some relationship of Mt. Wrangell with other volcanos or ranges such as the Truuli Ridge [Kenai Mountains], the Yakutat Mountains, or the Sea Alps of America's northwest coast as a whole. As we shall see below, there is reason enough to count the Near Islands and the Komandorskie Islands among the inactive volcanic islands, too. Therefore, it can hardly be doubted that from the Near Islands to the High Mountain there is one single row of volcanos, while it is not improbable that one single subterranean canal or strand of fire existed or still exists, reaching all the way from the east side of Kamchatka, past Kamchatskoi Nos, the Komandorskie, and the Aleutian Islands, and then by way of Alaska to Mt. Wrangell, the vents of which . form one row of volcanos. Southward from this canal we find the next fire-mountains known to us.[103] These are Mt. Baker (48°48′ N. Lat., 122° W. Long.) and other volcanos, which follow farther to the south. Westward, meanwhile, on Asia's coast the Kamchatkan row of volcanos is an extension of those on the Kuril Islands, or vice versa, between 52°42′ and 59°50′ N. Lat., 160° E. Long. The southernmost volcano of Kamchatka, the Kambalinaia Sopka,[104] falls approximately into the latitude of Amchitka Island, the southernmost of the Aleutian Islands. And the most prominent elevation of Kamchatka's row of volcanos, the Kliuchevskaia Sopka (56°04′ N. Lat.) is located in the latitude of Mt. Calder on Prince of Wales Island and of Veniaminof Volcano. Hoff maintains that Kamchatka's row of volcanos begins only in that location and continues southward from there to where the row of Aleutian Islands intersects with it by way of Bering Island. It ought to be added here that from earliest times volcanic activity has been known to occur farther northward still, under 59°50′ N. Lat., near Tumlat Village on Kamchatka (cf., Berghaus, *Geography and Ethnography* II, pp. 733–735, and Erman's *Reise um die Erde*, I, 3, p. 376). This is the approximate latitude of St. Matthew's Island and of Mt. Iliamna and Mt. St. Elias. There is also the northernmost smoking andesite elevation on Kamchatka, Mt. Shivelich, located one degree north of Bering Island. It forms a ridge, which sweeps NE to SW and runs contrary to the major range of Bering Island. The Kronotskaia Volcano and the Shchepinskaia Volcano are at that latitude, also the Milkover Khrebet of the same system of elevations. There are unmistakable interrelationships in the order and volcanicity of the single row of volcanos on the Aleutian Islands and Alaska, and between the double rows of Kamchatka and North America. This gives us cause to assume a subterranean connection, occasionally if not constantly, of those three rows of volcanos just mentioned. This would serve as proof that the entire Pacific Ocean is ringed with a belt or canal of volcanos, kept alive by that connection.[105] And the recent discovery of Mt. Erebus closed that ring, so to speak.

Inasmuch as longer dormancy of volcanos results in more violent eruptions, a larger distance of vents upon one and the same stream of fire might cause a heightened outward activity. Volcanos behave like vents on a steam kettle. And inasmuch as meteorology teaches an equalization of varying conditions of our atmosphere over a larger portion of time and space, so, too, could similar conditions prevail in the process within the earth's interior. Whenever blockages or disturbances occur within that mighty subterranean canal which surrounds the Great Ocean, then it seems that in other locations a heightened activity can be witnessed, as even the sparse data show, which we have from this belt. This, of course, is measured relative to the intensity of volcanic action over a certain time period.

That part of the cycloidally winding trench on which the Aleutian Islands had arisen, seems to have diminshed its activity since the time it had become known. The same has not been observed during the same period of time along the coast range of North America, i.e., from Mt. St. Elias southward to a latitude of 50°; although Buch (*Canary Isl.*, p. 390) assumes that the highest points of the same are "most likely" volcanos. We have thoroughly searched all known travelogues

(Appendix II): Bering (Steller in Pallas, *N. B.* V and VI), Cook (*Third Voyage*, by G. Forster, II, p. 67), Maurelle (Arteaga's Journey, in Marchand's *Voyage*, Vol. I, p. lxviii), La Pérouse (II, p. 219), Vancouver (III, p. 204), Roquefeuil (II, pp. 9, 17), Malaspina (Humboldt's *Nouv. Esp.* pt. III, Vol. I, pp. 399, 238, Vol. II, p. 487, pt. I, Physique Generale, p. 148, and Krusenstern, *Beitr. Hydr.*, 1819, p. 227), Belcher (I, p. 80), etc. From these, however, we only learn that they do not mention any volcanicity of Mt. St. Elias, Mt. Fairweather, Crillon, and Edgecumbe. Even the activity of the latter must be questioned from the time it has been known, as we attempted to prove in a footnote [endnote 20]. According to the reports of the gentlemen von Wrangell, Etolin, Kashevarov, Blaschke, Voznesenskii, and others, no one has seen the just-mentioned mountain smoking. Nor have any volcanic phenomena been observed on [Prince of] Wales Island. Thus, Maurelle's report (cf. p. 5 and Appendix II, 1775) of volcanos in the region of Bucarelli Harbor on Wales Island is unsure and probably erroneous, much like the error La Pérouse and Vancouver made with regard to Cape Mendocino. Nor were earthquakes observed by any travelers along this part of the coast. They only assumed that earthquakes had happened, because of seemingly recent changes in the surface of the earth (e.g., Roquefeuil, I, pp. 195–197: On the Island of San Miguel and L'anse Des Amis at Nootka Sound), or from the frequently most bizarre groupings of volcanic rocks.

Presently active volcanos in the Columbia Range, about which we know, are Mt. Baker, Mt. Rainier, Mt. St. Helens, and a mountain in the vicinity of Mt. Hood. But the actual Coast Range from Olympia via Cape Mendocino down the entire California Peninsula, reveals not a single point where exactly verified volcanic phenomena exist, since the time the region has been known. La Pérouse and Vancouver (III, p. 345, third of November, 1794) noticed fire and a column of smoke near Cape Mendocino, which they ascribed to a volcano. But Roquefeuil proved this opinion wrong (I, pp. xxvii–xl and II, pp. 227–229, and 244). He proved that the same phenomena, observed by himself in September of 1818, must be ascribed to burning grass, much like the grass fires on the prairie, which are usually set by the Indians in fall, so as better to be able to pick the wild peas which they eat. Likewise, the volcanicity of the Cerro de la Giganta and of [Cerro] de Los Virgines is not confirmed (cf. Proceedings of the Mineralogical Society at St. Petersburg, 1846, p. 143); the latter, at any rate, has not been active since 1746.

It follows from these deliberations that the mountains St. Elias, Fairweather, Crillon, Edgecumbe, Calder, Tasche (on Quadra, Roquefeuil, I, p. 210), and the mountains from Olympia to Cerro de Los Virgines have not been active since the middle of the last century [i.e., the eighteenth century]. For some of them, the point in time has to be fixed further back yet. Others have never been active. Generally, however, the law is verified, according to which the frequency of eruptions stands in an inverse relationship to the elevation of the volcanos. Thus, a long period of quiescence must be assumed for Mt. St. Elias, with 16,758 ft. of elevation above sea level the tallest—not absolutely but specifically—volcano. It overtops the Kliuchevskaia Sopka by 2882 ft. How frightful and disastrous would be the floods alone from the melting ice-covering, at the awakening of that activity!

We believe, however, that we have reason not to entertain that fear, although Mt. St. Elias rises along one line of volcanos and in the same latitude as Mt. Iliamna, and Nunivak and St. Matthew's Islands. The volcanicity can properly be doubted, because of its lack of conical shape and craters. And since its ridge terminates in many peaks, it is very probable that an active fire-vent or flue never did penetrate the mountain. Should it be argued, however, that among the jagged peaks of the long ridge of Mt. Pichincha no volcano was expected either, then we should always assume an exception from the common rule. Furthermore, the not too great distance that divides the still-active Mt. Wrangell from Mt. St. Elias, allows us to assume that the subterranean canal will not soon seek just that, perhaps very difficult, exit through the giant of North America's mountains. Another argument for the present and past volcanic activity of Mt. St. Elias can be found in Belcher's edition, which speaks of black stripes on top of the snow that covers the peaks of Mt. St. Elias, as well as vegetation that grows on top of the glaciers along the Copper River (Wrangell). The former report could be based on an optical illusion; and the black stripes do not have to be caused by volcanic ashes of all things. The latter phenomenon (the growth of vegetation on top of the glaciers) should be explained as an outfall of ashes that happened a considerable time ago, which covered the glaciers. This has been observed on Mt. Aetna, where in 1787 even a stream of lava flowed over the top of a layer of ashes that covered the ice. But the outfall of ashes could well have been caused by the still-active, although considerably distant, Wrangell Volcano. But this and the glaciers along the Copper River[106] did not become known until the beginning of our century (1819). The latter must have been covered with ashes before that time, so as to clothe themselves within seven years with almost tree-like vegetation. It always remains possible that the covering of the glaciers are moraines of a very high age. Or perhaps the observations of these glaciers are based on the same kind of error as were those of Eschscholtz's ice-formations at Kotzebue Sound.

We readily leave this field of hypothesis and turn now to a survey of all the volcanic phenomena of these regions in a geographic as well as chronological order, notwithstanding the sparse historical information, which must yet expand significantly. All ships' logs, notably the Russian ones, are in agreement on this. For a further expansion of the same, or a comparative survey of the phenomena along the volcanic belt surrounding the Great Ocean, as well as all the volcanic phenomena known to us from around the entire globe, we hope to publish in one of the next issues of this journal, after the samples dispatched by Voznesenskii from Kamchatka have been processed. It should furthermore be remembered that the data depended upon chance visits of European travelers. Only a few traditions could be collected, because of the low level of education of most of the natives. And finally, because of the frequent total silence, faint stirring, or very minor reawakening of volcanic activity, a comprehensive picture of these phenomena can be provided only as the result of constant thorough observations and recordings.

On Map 5, added here, there is a border between active and dormant volcanos, not too strictly drawn. The mountains marked with a red dot are those of which volcanic phenomena have been reported. Therefore, they must include locations where volcanic eruptions have taken place.

Volcanic Phenomena on the Northwest Coast of America and on the Adjacent Islands, in Geographic Order

Wales or Beaver Island

With Mt. Calder in 56°15' N. Lat., 133°49' W. Long.: 1775, active? (Don Antonio Maurelle); 1793, dormant (Vancouver); 1833, dormant (Etolin).

Sitka or Baranof Island

Hot springs, 56°51' N. Lat., 135°19' W. Long.: unchanged since 1779 (Baranov).

Kruzof Island

Mt. Edgecumbe, 57°03' N. Lat., 135°40' W. Long.: 1775, dormant (Antonio Maurelle); 1778, the same (Cook); 1786, the same (Dixon); 1791, the same (Marchand); 1796, active? (Hofmann). 1804, dormant (Lisianskii); 1818, the same (Roquefeuil and Golovnin); 1827, the same (Postels and Lütke); 1837, the same (Belcher); 1842, the same (Simpson).

Mainland of North America

Crillon, 58°45' N. Lat., 137° W. Long.

Mt. Fairweather, 59° N. Lat., 137°30' W. Long.

Mt. St. Elias, 60°17'30" N. Lat., 140°51' W. Long.: 1778, dormant (Cook); 1786–88, the same (Vancouver), until 1847.

Coastal volcano at Chugach Bay [Prince William Sound], 60°54' N. Lat.: eruptions in 1790 (Don Fidalgo).

Wrangell Volcano, 62° N. Lat., 142°–143° W. Long.: discovered in 1819; active, and several earthquakes annually (Klimovskii and Wrangell).

High Mountain [Redoubt Volcano], 60°30′ N. Lat., 152°45′ W. Long.: since it became known in 1819, it has been smoking (Wrangell).

Iliamna Volcano, 60° N. Lat., 153°15′ W. Long.: dormant (?) in 1741 (Bering); 1778, active (Cook); 1779, the same (Arteaga); 1843, the same (Voznesenskii).

Alaska Peninsula

Veniaminof Volcano, 56° N. Lat., 158°–159° W. Long.: 1830–1840, smoking (Veniaminov).

Hot Springs at Perenosnyi [Balboa] Bay, 55°35′ N. Lat., 160°27′ W. Long.: the same [*sic*] (Veniaminov).

Pavlovskii Volcano, 55°24′ N. Lat., 161°48′ W. Long.: 1762–1786, active (reported by promyshlenniks); 1786, one of the craters closed up (Chamisso); 1790, active (Sarychev); and smoking until the present time (Voznesenskii).

Medvenikovsky [Medvednikovskii] Volcano, 55°05′ N. Lat., 162° W. Long., and Morshovsky [Morzhovoi, North Walrus Peak] Volcano, 55°05′ N. Lat., 162°37′ W. Long.: 1768–1769, both dormant (Krenitsyn); 1790, Medvenikovsky(?) Volcano active (Sarychev). Later, dormant.

Hot spring on a small island at the entrance to Morshovsky [Morzhovoi] Bay, 54°54′ N. Lat., 152°50′ W. Long., possibly Amagat [Island]: 1832 (Voronkovskii, Lütke).

Hot spring at the bay of the same name: Goriachikh Kliuchei, 55° N. Lat., 163°10′ W. Long., 1838 (Veniaminov).

Hot spring at Moller Bay, 55°45′ N. Lat., 160°30′ W. Long.: 1828 (Lütke), 1840 (Veniaminov).

Volcanic island, Amak, 55°26′ N. Lat, 163°15′ W. Long.: active crater during the last century; dormant since 1804 (Krusenstern).

Unimak

Khaginak Volcano, 54°45′ N. Lat., 163°45′ W. Long.

Shishaldin Volcano, 54°45′ N. Lat., 164° W. Long.

Pogromnaia Volcano, 54°30′ N. Lat., 164°45′ W. Long., with hot springs and swamps. 1690, crater formation on the highest mountain east of the Shishaldin (according to Veniaminov). 1775–1778, a volcano on Unimak Island (most likely Mt. Shishaldin), frequently on fire (Zaikov). 1778, Shishaldin smoking (Cook). 1790, Shishaldin smoking (Sauer), until 1825 (Veniaminov). 1795, the SW end of the island burst open; outfall of ashes, and newly awakened activity on the NE and NW sides of Pogromnaia Volcano, while another volcano on the northern side of the same is extinguished (Veniaminov). Toward the end of 1821 and until the tenth of March of 1825 the Shishaldin belches fire (Veniaminov). 1825, tenth of March, volcanic eruptions in the Isanak Range on the northeast side of the island (Lütke, Veniaminov). 1825 (March) until 1827 (March), the Shishaldin smokes (Veniaminov). 1826, renewed eruption in the island's interior, not far from the southern end; outfall of ashes (Veniaminov). 1827 Pogromnaia belching fire (?) (Isenbeck). From March 1827 to 1829 the Shishaldin (?) belching fire (Veniaminov), then smoking until fall of 1830. 1830–1831, Shishaldin particularly active (Lütke), spraying of flames, and glowing fissures, later on smoking (Veniaminov).

Akun

Akun Volcano, in 54°17′ N. Lat., 165°33′ W. Long. Hot springs on a small island on the NW side of Akun (Postels, 1828, and Veniaminov). 1765–1770, as this and the next following island were discovered, no volcanic phenomenon was observed. 1828, smoke columns periodically expelled (Postels).

Akutan

Akutan Volcano, 54°08′ N. Lat., 165°54′ W. Long. Hot springs. 1778, dormant (Cook); 1785, the same (Shelikhov); 1790, smoking (Sauer and Sarychev); 1828, smoking and hot springs (Postels).

Unalaska

Makushin Volcano, 53°52′ N. Lat., 166°48′ W. Long. Hot springs. 1768, two active volcanos (Krenitsyn); 1778, dormant? (Cook). 1790–1792, two burnt-out volcanos (Sauer) and smoking, but not belching any fire (Sarychev). 1802, strong earthquake and eruption (?) of the Makushin, while St. John Bogoslof is inactive (Langsdorff). 1816 and 1817, no smoke (Eschscholtz). 1818, earthquake near the Makushin (Veniaminov). 1826, Makushin erupts in fire (?), two strong earthquakes, hot springs (Postels). 1843 and 1844, smoking (Voznesenskii).

Umnak

Tulikskoi Volcano, 53°20′ N. Lat., 167°50′ W. Long. Sevidovskii [Vsevidof] Volcano, 53°10′ N. Lat., 168°12′ W. Long. Recheshnoi Volcano, 53° N. Lat. (?), 168°24′ W. Long. Hot springs at the Sevidovskii Volcano.

1765–1770, no volcanic phenomena reported. 1784, Sevidovskii smokes; and there are hot springs at its base. 1790, Sevidovskii is smoking (Sarychev). 1817 (or 1820), terrible eruption and expulsion of ashes at the northern end of the island, perhaps from Tulikskoi Volcano. Elevation of a strip of land (Lütke and Postels). 1828, hot springs appear northeast of the Sevidovskii Volcano. 1830, at the SW end of the island renewed activity of a volcanic cone, perhaps the Recheshnoi (Veniaminov).

Volcanic Island St. John Bogoslof

Volcano in 53°56′20″ N. Lat., 167°57′ W. Long. Hot springs. Came into existence 1796, the seventh of May; earthquakes and eruptions (Krusenstern and Langsdorff). 1800, no smoke (Kotzebue); 1802, the same (Langsdorff). 1804, steam from one crater (Kotzebue). 1806, on fire on its northern side; lava stream (Langsdorff). 1814, the crater expels rocks (Baranov). 1815, diminishing in elevation (Baranov). 1816 and 1817, no activity (Eschscholtz). 1820, smoking (Dr. Stein). 1823, no smoke (Veniaminov). 1832, no smoke (Teben'kov and Lütke).

Kigamiliakh [Kagamil Island]

Volcano at 52°53′ N. Lat., 169°30′ W. Long. Hot springs. Formerly eruptions (according to Lütke and Postels). Later dormant, but in 1828 hot springs, steam exhalation, and subterranean rumbling on the southern side.

Tanakh-Angunakh [Chuginadak Island]

Volcano at 53° N. Lat., 169°45′ W. Long. Hot springs. Before 1774 (Bragin), supposedly consisting of two islands and dormant. Later, 1828, an active volcano, and hot springs at its base (Lütke).

Ulaegan and Chegulakh [Carlisle and Herbert]

Volcano at 52°53′ N. Lat., 169°40′ W. Long. and 53°08′ N. Lat., 169°24′ W. Long. Since the eighteenth century, when they became known, the volcanic cones have been dormant. Once, they had supposedly been active (Voznesenskii).

Yunaska

Volcano at 52°40′ N. Lat., 170°28′ W. Long.: 1817, in April, smoking (Choris). 1824, renewed eruption of a volcano (Lütke). 1830, explusion of ashes.

Amukta

Volcano at 52°30′ N. Lat., 171°04′ W. Long.: 1786, in June, all in flames (Shelikhov). 1790, active (Sarychev). Later reports are missing until 1830, when it was quiet (Lütke).

Siguam or Seguam

Volcano at 52°20′ N. Lat., 172°12′ W. Long. Mud craters and hot springs. Active prior to 1790 (Sarychev); 1827, smoking (Lütke). Mud craters and hot springs (Lütke).

Atka

Sarichef Volcano in 52°20′ N. Lat.(?), 173°47′ W. Long. Kliuchevskoi Volcano, 52°20′ N. Lat., 173°55′ W. Long. Korovinskoi Volcano, 52°24′ N. Lat., 173°55′ W. Long. The Conical Sopka, 52°22′30″ N. Lat., 174°06′ W. Long. The Sergief Sopka, 52°18′ N. Lat., 174°09′ W. Long. Mud craters and hot springs.

Since 1760, several volcanos alternatingly active (Baikov, Tolstykh, Lütke). The Sarichef Volcano no longer on fire since long before 1792 (Sarychev), but in 1812 strong earthquakes and eruptions (Golovnin after Vasil'ev). 1829 and 1830, the Korovinskoi Volcano is smoking (Ingenstrom). Mud craters and hot springs near the Conical Sopka, 1827 (Lütke), 1829 (Ingenstrom), 1844 (Voznesenskii). Hot springs near the Kliuchevskoi Volcano (the same authors). 1844, weak smoke from the Korovinskoi Volcano. The rest dormant.

Koniuzhii Islands [Koniuji]

Volcano at 52°13′ N. Lat., 174°54′ W. Long. Supposedly in the process of constant elevation, and in 1827 smoking in the center (Lütke).

Kasatochi

Volcano at 52°09′ N. Lat., 175°14′ W. Long.: 1827, extinguished crater (Lütke).

Great Sitkin

Volcano at 52°04′ N. Lat., 176°02′ W. Long.: 1760, quiet (Tolstykh); 1792, belching fire (Sarychev); 1829, covered with snow and smoking (?) (Ingenstrom).

Adak

The White Volcano (? 51°40′ N. Lat., 176°30′ W. Long.). Hot springs. 1760, active (Tolstykh, Baikov); 1784, active, hot springs (Shelikhov); 1790 and 1791, no reports on its activity (Sauer and Sarychev).

Kanaga

Volcano at 52° N. Lat., 176°50′ W. Long. Hot springs. 1763, volcano with crater where sulfur was collected (Solfatara), and hot springs (Tolstykh). 1786, erupting with fire (Shelikhov). 1790 and 1791, smoke rising from the hot springs at the base of a once-active volcano (Sauer, Sarychev). 1827, smoking (Lütke).

Tanaga

Volcano at 52° N. Lat., 178° W. Long.: 1763–1770 constantly active (according to promyshlenniks); 1791, smoke observed (Sauer).

Gareloi Island

Volcano at 51°43′ N. Lat., 178°45′ W. Long.: 1760, active (Baikov); 1792, belching fire (Sarychev); 1792, smoking (Ingenstrom).

Semisopochnoi

Volcano at 52° N. Lat., 179°45′ E. Long.: 1772, smoking (Bragin); 1790 and 1792, the same (Sauer and Sarychev); 1830, the same (Lütke).

Sitignak

Volcano at 51°43′ N. Lat., 178°22′ E. Long. or 51°39′ N. Lat., 178°27′ E. Long. Hot springs. 1776, fire-belching mountain (Bragin).

Little Sitkin

Volcano in 52° N. Lat., 178°30′ E. Long.: Smoking 1828 (Lütke).

Chronological Survey of the Volcanic Phenomena on the Aleutian Islands and on the Northwest Coast of North America

1690 Crater-formation on the highest mountain east of Mt. Shishaldin on Unimak Island; perhaps Mt. Khaginak.

1700–1710(?) Volcanic activity on Ulaegan, Chegulakh, and Amak (beginning of the eighthteenth century).

1741 Mt. Iliamna, or Bering's Mt. Dolmat (?) quiet.

1760 Adak, Gareloi, Chetchina, and Atka smoking to the present. Koniushi [Koni-uji] Island is rising.

1762 Pavlovskii Volcano on Alaska active.

1763 Tanaga active until 1770. Kanaga Solfatara.

1768 The Aiagish and another volcano on Unalaska active, on Alaska the Medvenikovsky and Morshovsky.

1770 Amukta, with active volcano until this year.

1772 Semisopochnoi smoking.

1774 Tanakh-Angunakh active.

1775 Mt. Calder and several neighboring mountains on Wales Island supposedly active. Until 1778 a volcano on Unimak occasionally on fire.

1776 In July Mt. Sitignak belching fire.

1778 Mt. Iliamna active, up to the present; and Shishaldin on Unimak smoking.

1784 Sevidovskii [Vsevidof] on Unimak smoking; also Chetshina, in July.

1786 Kanaga erupting fire. Pavlovskii Volcano active. Until 1790 Seguam, and until 1791 Amukta active.

1788 Reports of particular volcanic phenomena are absent. Noteworthy is the flood or deluge which occurred on July 27 in Sanak, Unga, and parts of Alaska.

1790 Akutan smoking, as well as Umnak (Sevidovskii), Kanaga, and Semisopochnoi. Makushin on Unalaska, active between 1790 and 1792, and Mt. Agayedan [Shishaldin] on Unimak alternate actively between 1790 and 1825. Eruption at Chugatsk Bay in 60°54′ N. Lat.

1791 Tanaga and Kanaga smoking (June).

1792 Great Sitkin and Gareloi belching fire until the end of May. Semisopochnoi smoking on the first of June.

1795 Eruption and expulsion of ashes on the SW end of Unimak; simultaneous extinction of heretofore active volcano on the north side of the Pogromnoi.

1796 Seventh of May: genesis of St. J. Bogoslof. (Edgcumbe active?)

1796–1800 Four-Cone Islands [Islands of Four Mountains] active at the end of the eighthteenth century. Amak with active crater (?).

1800 Until 1815, St. J. Bogoslof in the process of growing, but no smoke is expelled.

1802 Makushin, strong fire-eruption (?), earthquake. St. J. Bogoslof quiet.

1812 Sarychev Volcano on Atka active after long dormancy; strong earthquakes.

1817 The first of March, eruption, falling of ashes, earthquake and SW storm at the north end of Unimak, Yunaska smoking at the beginning of April.

1818 Earthquakes at the Makushin on Unalaska; and on Amaknak reportedly large alterations.

1819 Wrangell Volcano erupting in fire, and the High Mountain [Redoubt Volcano] smokes.

1820 St. J. Bogoslof smoking.

1824 Shishaldin on Unimak erupts with fire from the end of 1824 until the tenth of March, 1825. On Yunaska, an enormous eruption after long dormancy.

1825 Volcanic eruptions on Unimak's east side in the Isannak Range on the tenth of March.

1826 On Unimak's southern point eruptions and falling of ashes (eleventh of October). The Makushin of Unalaska smokes; and in June there are two earthquakes.

1827 Eruption of Shishaldin and Pogromnoi on Unimak, March 1827 to 1829. Koniuji and Kanaga smoking. In June an earthquake on Copper Island.

1828 Little Sitkin, Akun, Akutan, Tanakh-Angunakh, Atka, Koniushi [Koniuji], Gareloi, and Unimak (Shishaldin) smoking.

1829 On Unimak, Shishaldin smokes until fall of 1830. Great Sitkin, Gareloi, Tanaga, Kanaga, and Atka also smoking.

1830 Until 1831, Shishaldin erupting with fire, and a new eruption at the SW end of Umnak in August. On Yunaska eruption of ashes. Korovinskoi on Atka is smoking.

1836 Three smoking locations on Unimak. Shishaldin erupting with fire. Tanakh-Angunakh, Makushin on Unalaska, Akutan, Pavlovskii Volcano, and a mountain range on Alaska in 56° N. Lat., 158–159° W. Long. (Veniaminof Volcano) smoking.

1843 Eruption of St. Helens (twenty-third of November, New Style.)

1844 Korovinskoi on Atka and Makushin on Unalaska weakly smoking.

From this, certainly incomplete, survey, the fact seems to emerge that the volcanic activity of the Aleutian Islands and of Alaska is in the process of subsiding since the time the islands first became known. Presently the major terminal locations of the northern part of that subterranean canal surrounding the Great Ocean are on Kamchatka the Kliuchevskaia Sopka, and on America's mainland the Wrangell Volcano. On the island chain between Asia and America they are found in the group of the Fox Islands. There are three manifestations of volcanic activity: eruption, solfatara, and complete dormancy. If we apply these to the above-mentioned twenty-five volcanic islands, then we see about the year 1830 solfatara on twelve of them (Little Sitkin, Semisopochnoi, Goreloi, Tanaga, Kanaga, Koniushi, Atka, Seguam, Kigamiliakh, St. J. Bogoslof, Akutan, Akun). Eight islands were completely inactive (Sitignak, Adakh, Great Sitkin, Kasatochi, Amukta, Ulaegan, Chegulakh, Amak). And five were in definite, if not always strong or continuing action (Yunaska, Tanakh-Angunakh, Umnak, Unalaska, Unimak).

It is furthermore not to be denied, that there are certain interrelationships in existence between dormancy and activity of points of various distance from each other along the large northern volcanic belt. According to the oldest reports from Tolstykh, Baikov, Bragin, Zaikov, Shelikhov, Cook, Sauer, Vancouver, etc., the islands Sitignak, Kanaga, Amukta, Kigamiliakh, St. J. Bogoslof, Unalaska, Unimak, the volcanos of Alaska, and Mt. Iliamna were in intermittent, but diminishing activity from the middle toward the end of the past [eighteenth] century. At the same time the main seat of volcanic activity was progressing from west to east. On Kamchatka (where between 1727 and 1731 the Kliuchevskaia Sopka was continuously on fire; 1737 the same, and the Avachinskaia Sopka had enormous eruptions, and the Tolbachinskaia became notorious in 1739) we hear during the second half of the eighteenth century of only two large eruptions of the Kliuchevskaia Sopka, in 1762 and 1767,[107] of the Avachinskaia Sopka in 1773, and of the powerful eruptions of the Kambalinaia and Kliuchevskaia Sopkas at the end of that century, in 1796. Thereafter, further reports are wanting. And it is possible that no particularly conspicuous phenomena occurred, but rather that comparative quiescence set in until 1820, when the fire-stacks of Yunaska, Umnak, and Umimak vigorously came to life, while the activity of the Wrangell Volcano also became notorious. But when after this point in time the intensity of volcanic powers on the islands just mentioned began to subside, two volcanos on Kamchatka opened their funnels: Avachinskaia Sopka in 1827, and Kliuchevskaia Sopka in 1829. In recent time we have received no news about any significant volcanic phenomenon from the Aleutian Islands or the adjacent mainland of North America. But on Kamchatka, the Kliuchevskaia Sopka has begun its activity again, in 1848, after six years of dormancy.

Although these indications do not amount to evidence of an interconnection of subterranean canals progressing in different directions, there is support for it nonetheless, in that we see in smaller regions, such as the islands Umnak, Unalaska, and Unimak, the activity of a cinder cone subsiding, while another one awakens to renewed activity.[108] Whether the same kind of interrelationship could also explain the dormancy of Mt. Edgecumbe, on the same latitude as Kodiak, Mt. Chiginagak on Alaska, and the Pribilof Islands, after the emergence of St. J. Bogoslof, this is hard to decide. The reports on the last activity of Mt. Edgecumbe are dubious; and the reports on the vast distances in between are most inadequate. Nevertheless, it becomes inadvertently apparent that there is an interconnection among these phenomena; and such reports gain in probability when the widely distributed spectacular activities of volcanic powers in 1795, 1796 and 1797 are remembered.[109] Then the South and Middle American volcanos had significant eruptions; also reportedly, Mt. Edgecumbe, and furthermore Unimak (SW end), Umnak (N end and St. J. Bogoslof), Kamchatka (Kambalinaia and Kliuchevskaia Sopkas), the Kuril Islands, etc. And during the year 1796 it was reported that Mt. Tashen in Java, and perhaps Mt. Edgecumbe had gone dormant. But as we mentioned above, it is not our intention to continue similar speculations in this place. We only want to mention in passing that Mt. St. Helens, for example, erupted significantly in 1848, the year the Kliuchevskaia Sopka awakened to renewed activity. Or if we hear (what we probably will hear) that after an extended activity it began to quiet again, it might serve as proof that these two volcanic funnels, so far distant from each other, might be related with each other.

The row of Aleutian Islands, with the Alaska Peninsula in the east, and the Komandorskie Islands in the west, forms a bow-shaped line, which is strung out like a knotted rope or a chain between the rock columns, America and Asia, as if it had sunk under its own weight, thereby pulling its anchoring-points toward each other. Three mighty mountain ranges, the Rocky Mountains,[110] the Columbia Mountains, and the Sea Alps of the NW coast rise in fairly parallel extension from SE to NW toward that arch. This, in turn, connects to the mainland from SW toward NE by way of the Alaskan mountain ranges, as well as the Truuli [Kenai] and the Chigmit Mountains. On Asia's coast, the middle mountains of Kamchatka form two parallel ranges from SSW to NNE. These are approached by the Komandorskie Islands in a SE to NW extension, consistent with the peninsula between the Kamchatskii and the Stolbovoi Nos. When we observe the shape of the larger islands between Asia and America, then we notice that Bering Island and Copper Island extend SE to NW, Attu from W to E, Amchitka in both directions just mentioned, Tanaga from W to E,[111] and almost all of those larger islands located east of Tanaga[112] from SW to NE or WSW to ENE. Most of the latter islands, except for Unimak, are lower on the W and SW side and higher on their N and NE sides.

Among these elevation directions, indicated by the directions of the mountains, earthquakes, etc., the SW to NE direction has been the most prevalent and most pronounced of all. The latest elevations in the Fox Islands are also predominantly restricted to that direction, although they are observed on the northern side of the just-mentioned islands only (Veniaminov, I, p. 7). That direction goes through WSW to ENE over to W to E, and expresses itself in the entire extension of the Aleutian chain of islands. The latter direction, from W to E, is especially conspicuous in the linear position of the volcanos on the fifty-second degree of latitude. On Amchitka (the eastern half of which retains the W to E direction; but its western half extends SE to NW), Aguyadakh, the Semichi Islands, and especially the Commander Islands the contrary main direction, SE to NW, is predominant. It is the one which characterizes the largest part of the American continent.

Thus far have we been led in our considerations of the configurations of these islands. But, never did the glance of a trained geologic observer go over them. And for that reason we lack indications concerning their shape, which are absolutely essential as proofs for a theory of elevation islands and elevation craters. Except for geologic combinations, no report and no illustration lets us assume any elevation craters with cones of trachyte or andesite rising from them, except for the Commander Islands and Little Alaid. We therefore must restrict ourselves to assume the existence of twenty-five true volcanos. Still, we believe that among these there are a few elevation islands, and among the other mountains quite a few more elevation islands in existence, where the emerging volcanic core has not broken through the active state; or maybe only a few eruptions took place. All three forms of islands, which according to von Buch's opinion indicate elevation islands, i.e., eruption, volcanic, and basaltic forms, seem to be present among the islands between Asia and America. The following remarks are to affirm these assumptions. And on the basis of the material on hand, they are to provide a corresponding overview of the geologic conditions. They should make them more tangible, and on the whole they should show that in the chain of the Aleutian Islands there are volcanic, as well as eruption and basaltic islands.

1. Except for the Shishaldin Volcano, all the higher and reportedly active volcanos are covered with permanent snow at the top.

2. The position of the craters has been learned on the following island volcanos:

Shishaldin (Unimak): at the summit of the pointed cone.
Khaginak (Unimak): at the summit, which was torn apart and has collapsed.
Akun: at the top of a mountain peak.
Akutan: the old crater at the top of a volcanic cone, a new crater at the north side.

Makushin (Unalaska): at the summit of the blunted cone.

Sevidovskii (Umnak): on the ridge-shaped summit.

St. John Bogoslof: on top of the pyramid-shaped island.

Chegulakh: at the top of the cone.

Yunaska: summit crater of the blunted cone.

Korovinskoi (Atka): two peaks with depressed ridge; the crater possibly in that place.

Kasatochi: on top of a rounded mountain ridge an old crater (lake).

Kanaga: at the top of the cone.

Tanaga: at the summit of one of the many peaks of the major elevation.

Little Sitkin: side-crater of a cone-shaped mountain.

Most of the craters of these volcanos, which occasionally are of significant elevation,[113] are thus located at their summits. About the volcanos on the mainland we know that Medvenikovsky [Medvednikovskii] and Pavlovskii on Alaska have collapsed summits. Mt. Iliamna, and perhaps Wrangell, too, have sidecraters. And Edgecumbe, finally, has a summit-crater. As we now consider the position of these craters, we must assume that those at the summits belong to true volcanos. But the reports we have to date do not seem to prove adequately that those orifices from which steam is escaping are also true volcanic craters which indicate a communication of the earth's interior with the atmosphere for an extended duration.

3. The cause of the smoking of several island volcanos can be traced to hot springs, as on Kanaga (Sauer), or to steam exhalations as on Akutan (Veniaminov). The steam escapes either from fissures, which are essentially different from craters, or it is part of rare volcanic activity in elevation craters, or finally, it could be the remains of activity of true volcanos, the craters of which have become stopped up. But wherever there occurs a periodic expulsion of smoke, especially in autumn, as was observed on several of the Aleutian volcanos, there a constant communication with the volcanic hearth could be assumed. This observation on the part of the natives and of all the settlers who have stayed on Kamchatka for an extended time, and on the Aleutian Islands and Alaska, speaks against the usual assumption that the behavior of volcanos is absolutely independent of meteorological conditions or the seasons of the year. In autumn the atmospheric precipitations are extremely abundant in these regions. And in the surroundings of a volcano they need hardly penetrate into significant depths in order to be expelled through the main crater in the form of steam. But whether that water penetrates to the depth where the temperature of the earth's interior reaches the degree that melts lava, and whether this, in turn, is lifted up by the force of the resulting steam, is a question that needs not be answered here. The fire-belching of these volcanos,

and similar indistinct expressions for this phenomenon in other languages, might be flames of gases in some instances, and in others the reflections of those masses of glowing lava moving about in the crater's interior, but seldom true eruptions and expulsions. Thus Wrangell, too, says of the volcano named after him, without mentioning a crater on its side: "The volcano, which expels fire constantly, is covered with snow all the way to its summit."

4. Although different kinds of lava, obsidian and pumice stone are found on Alaska and the mainland beyond, as well as on the Aleutian Islands, we hear very little of any streams of lava observed there. And there is no reason to assume that such conspicuous phenomena might have been overlooked or the reporting of them neglected. Baranov and Stein mentioned a lava stream at St. J. Bogoslof (1804–1806), Lütke on the southern side of Unimak (1827, September). Then those glowing fissures which Father Veniaminov observed from a large distance on Mt. Shishaldin in December of 1830 might have been streams of lava, which had an angle of descent of 6°. And finally, there were supposed to have been lava streams which were observed on the Pavlof Volcano (Alaska), and on the Korovin Volcano (Atka), according to the reports of Mr. Kashevarov and Mr. Voznesenskii.

5. Thus, the few eruptions of the volcanos of this region, with the exceptions just mentioned, consisted of rains of ashes and expulsions of lapilli, which were accompanied by floods of mud. The white appearance of the ashes expulsed from the Pogromnoi Volcano, gives cause to assume an eruption, on the basis of experiences and observations made in Europe. Lapilli and volcanic ashes can be found on most of the islands. The floods of mud probably derive for the most part from flash floods due to melting snow and ice, and seldom to subterranean cisterns of water as on Mt. Carguairazo (on the high plateau of Quito, 1698). A large part of the eruptions observed on the Aleutian Islands did not occur from the main craters of volcanos, as could be observed especially on the many newly formed openings, cinder cones and craters on Unimak and Umnak. Instead, these eruptions were for the most part single subordinate volcanic activities.

6. We often hear of sulfur, which is collected from the craters and sometimes from the slopes of volcanos. But we know now that the formation of crystalline sulfur from the sulfuric exhalations is the result of the weakest kind of stirring volcanic life. Thus, most of the supposed Aleutian volcanos must be solfataras at this time (cf. above).

7. There are several reported phenomena, such as the cave-ins and collapse of mountains, the elevation of the coast of Alaska and of the east coast of the Bering Sea, the swelling and emergence of large parts of islands (Koniushi, Tanakh-Angunakh), and furthermore the bursting of the western end of Unimak with an enormous report, and finally the above-mentioned fissures, thought to be

streams of lava, as reported by Veniaminov at Mt. Shishaldin, and similar ones at Mt. Iliamna, which could be taken for barancos. All these circumstances lead us to assume, in the complete absence of reports of other, similar phenomena commonly accompanying volcanic eruptions, that here is witnessed a continuous process of the formation of elevation cones, calderons, and barancos on the islands as well as on the mainland. A continuous gradual elevation of the land is occurring on the north coast of Unimak and Alaska, as well as generally along the entire eastern coastal margin of the Bering Sea. And while the depth of the sea in the vicinity of the islands is usually 100 fathoms, the water along the just-mentioned region in the Bering Sea is very shallow. But since we lack more accurate knowledge of the phenomenon of these elevations and its limits, we cannot compare them either with the four elevation areas[114] of the Old World, nor with those in Chile.

8. The form and position of some island groups and bays seem to indicate that they are elevation islands or rims of craters, such as those in the Greek archipelago. But since other important data are wanting for such an assumption, we have to refrain from concluding this hypothesis. We refer, instead, to the map where examples of this kind can easily be found.

9. If we observe the appearance, growth, and diminishing of St. J. Bogoslof Island more closely, we notice that it possessed for only a short time a crater that produced expulsions, such as Neo-Kaimeni did. The island arose slowly; and it is not the sole result of the eruptions of a sub-oceanic volcano, nor of a cinder cone, such as Ferdinandea.[115] As proof for the latter assumption serves above all the fact that St. J. Bogoslof continues to withstand the ocean waves for half a century already, that the island has the shape of a pyramid, while the crater was never significant enough to warrant detailed descriptions of it on the part of its observers. Amak should be taken for a true deposition cone, according to reports. But this island might have emerged analogous to St. J.Bogoslof. Both of them, however, seem not insignificantly different with regard to their genesis, from the Pribilof Islands, St. Matthew, St. Michael, and Stewart [Stuart]. They exhibit more the character of eruption islands. Their craters did not establish any continuous connection of the interior with the atmosphere. Only singular volcanic eruptions took place there, which originally caused the emergence of those islands. On the whole the actual Aleutian Islands, so far as we know them at this time, lack clear indications for the assumptions of their formation as eruption islands or deposition cones. Nevertheless, part of the process of the emergence of the latter has perhaps not escaped observation, as on some of the perfectly conical mountains or volcanos of Attu, Simichy [Semichi], Little Sitkin, Semisopochnoi, Kanaga, Great Sitkin, Kasatochi, Atka, Amlia, Amukta, Chugul, Ulagan, Chegulakh, Tanakh-Angunakh, Kigamiliakh, Kigalga, and Unimak. But Buldir, Goreloi [Gareloi],

Tanaga, Koniushi [Koniuji], Atka (Korovinsky), Siguam [Seguam], Yunaska, St. J. Bogoslof, Umnak (Sevidovsky), Akun, Akutan, Mt. Khaginak on Unimak, Mt. Pavlovskii and Medvenikovsky, on the other hand, are covered with pointed rocks, which threaten to fall down or actually collapse. This is reminiscent of the volcanic high peaks of the Andes, which consist of enormous masses of andesite blocks, piled one on top of the other. And between them there exist empty spaces and enormous caves, and only exceptionally true craters with lava streaming from them. They serve as vents for the expulsion of volcanic steam. It follows from this analogy that wherever cinder cones and lava are absent, the trachyte or andesite of the Aleutian Islands must have emerged gradually in a firm condition. But wherever olivine-lava as well as feldspar-lava is found, there it is not too daring to assume a combination of formation process of eruption and elevation craters, from the midst of which arise true volcanic cones with constant eruption activity. And more detailed reports and examinations of the otherwise not very inviting summits and craters, those among them perhaps specifically suited for this, will convince us to end the controversy and the contradictions in the theories regarding elevation craters and true volcanos.

Before we end these remarks, we must change the assumption which seemed so natural at the outset, namely, that the Aleutian Islands have twenty-five true volcanos. Based on the outer appearance alone, not even regarding the geologic compositions, not all of them might be so. Some of them have the peculiar character of certain pseudo-volcanos of the Andes; and others are mere eruption islands. The emergence of true volcanos can presently perhaps best be observed on Mt. Shishaldin. And the difference in the craters and fissures can be seen at the whale-shaped hole of Akutan, the summit of Khaginak, and on the plateau of Mt. Makushin. Atka's northern end and especially the slope of the Pogromnaia Volcano would enlighten us about the nature of the volcanic orifices, which in these places surround the volcano. Upon these observations would have to be based the decision whether Akun, Akutan, and others, especially the Four-Cone Islands [Islands of the Four Mountains], are volcanic islands or eruption islands, and whether the last-named group have risen on two parallel fissures, or whether they are closely related to each other, or to a common central point. The structure of the islands does not lead to any assumption that the straits between them could be remnants of barancos. And it would contradict every experience, even if this were indicated, that eruptions could occur at the summit-rim of an elevation crater, inasmuch as in such a small area a circular fissure is hard to assume, but rather a wider hearth of volcanic activity, unable to break through at its center. Noteworthy, however, is that the northern side of Kigalga, the west side of Tanakh-Angunakh and the south side of Kigamiliakh were thought to be active. But this information is not enough to determine if one, or which one, of these islands can be seen as the central volcano. The character of row volcanos of these islands, like Bogoslof, is complex. Seemingly, true volcanos, andesitian pseudo-volcanos, and eruption cones occur on one and the same running fissure. A closer examination of all the islands between America and Asia will perhaps close many a volcanic cone and crater forever. Others will be converted into eruption cones with terminal craters for single volcanic eruptions. And some, finally, where the latter characteristics were not observed, will receive their due honor.

A more detailed examination of earthquakes and the regions affected by them is not yet possible. Some seismometer stations would be desirable: near Okhotsk, on Kamchatka, on one of the Aleutian Islands reports presently available restrict themselves to the fact that the earth tremors in the Fox Islands move in a direction from SW to NE, those on the Pribilof Islands from W to E. The hot springs, too, and the mud craters, gas exhalations, etc., provide for the time being hardly more substance for discussion than can be found in a description of the individual islands. Therefore, we shall continue with some general observations concerning the geologic composition or the distribution of the different kinds of rocks of this region.

But before this happens, we point out the sad circumstance that we can only utter assumptions concerning the stratification of the different kinds of rock, collected by Voznesenskii, and described by us.[116] Thus, wanting more detailed reports and observations on our part, we could not go so far afield with our reflections, comparisons, and results as, for instance, Dr. Erman was allowed to go in his description of the adjacent parts of Asia. We therefore searched for a criterion for a more circumstantial development of the geologic conditions of the archipelago between Asia and America, and of the NW coast of the latter. And we thought we had found it in the just-mentioned description of Kamchatka, etc. But we soon realized that, because of the uncertainty of these data, we had to limit ourselves to the main results of Erman's investigations in their most general relationship to the geology of the regions with which we are here concerned. It happened after we realized that the geologic suite collected by Mr. Voznesenskii at the end of his voyage, with more care and experience than before, as well as the older collections in the Academy of Sciences in St. Petersburg, from Steller to Postels, and the newer ones as well, received from Middendorff,[117] Stubendorff, and others from the vicinity and coast of Okhotsk, were important enough not to be rendered superfluous by Erman's travelogue[118] and his other proper works, but that they should rather result in a closer critique of the same.

A transection [cross section or traverse] from Yakutsk to Okhotsk produces the following results, according to Erman's geologic sketch of North Asia (Erman, *Archive*, Vol. II), according to an essay on the geologic conditions of North Asia (Erman, *Archive*, Vol. III), and according to his travelogue (I, 3, with map of Kamchatka, Berlin, 1838, S. Schropp & Co.):

On the Lena and Aldan Rivers: Jurassic deposits.

On the Belaia: graywacke-limestone.

The major mass of the Aldan Mountains: graywacke, sandstone, and clay-slate.

Along the Okhota River: feldspar porphyry.

And north of her upper run: greenstone and granite.

On the coast of Okhotsk, south of the Urak: graywacke-limestone.

In the Marekanean Range there first occur granites and diorites, then stratified deposits containing coal, futhermore melaphyres, and finally trachytes and feldspar-like vitrifications.

The layered, coal-containing rock ([Erman's] *Reise*, I, 3, p. 95) resembles closely the main lode of the Aldanian graywacke. And the transition layers, remnants of which are preserved on both sides of the Aldan range, were during their formation already most densely penetrated by the feldspar-like substance of the euritporphyry which had lifted it up, although it did not actually form the crystalline core of this system. Not until later, but probably simultaneously with the volcanic events on Kamchatka, did these deposits, meltable according to their origin, get sintered, because of completely plutonic masses which broke forth here along the coast. Parts were transformed into trachytes, and yet other portions, deposited in pockets, had completely melted.

In Kamchatka, according to the same author, the western part, which consists of step-like ledges along the coast, is at first overlain with a chalk formation[119] through which emerge outcroppings of feldspar-porphyry.[120] Then follows a parallel strip of Tertiary deposits,[121] which contain melaphyres. And finally there follow in the middle range the end products of extinct lava-volcanos.[122] They also extend in the shape of veins almost the entire length of the peninsula, that is, after they terminate north of Uka, and re-emerge at the Tamlat River, and continue southward through the Kuril Islands. The middle range unites, incidentally, in 54° N. Lat. at the source of the Kamchatka River, with the range of the eastern row of volcanos of this peninsula. And Erman believes (l.c., p. 413) to be able to deduce from shore ledges and diluvial formations, that between the mountains just mentioned, and the present bed of the Kamchatka River with its numerous tributaries, there exist the remnants of a lake, which extended from the mouth of the Yelovka River possibly as far as the Kronotskian Lake. Northward along the estuary of the Yelovka River there are found graywacke, aphanite, diorite-porphyries, and diorite or granite. From among the volcanos of the eastern volcanic range, Mt. Shivelich,[123] the Tolbachinskaia Sopka, the Kronotskaia Sopka,[124] and the Streloshnaia Sopka consist of andesite, which usually is found at the foot of the mountains, tightly enveloped inside augitic and labrador-containing types of lava. The Kliuchevskaia Sopka consists mainly of such augite-porphyries, dolerites and lavas, as does probably the Avachinskaia Sopka, and other volcanos as well. In the Milkov and Ganalian ranges these types of rock from the eastern row of volcanos are covered with crystalline and shale-like rocks, which had their genesis in the graywacke period (p. 495). They seem to extend as far as the mouth of the Iupanova River (pp. 469 and 557). Thereupon there follow along the Bystraia and Bol'shaia rivers modified transition deposits (coarse shaly aphanite, greenstone). On the left bank of the first-named river, the sulfur springs of Malka emerge from not exactly defined deposits, probably volcanic masses (pp. 502–506). There are, moreover, four other similar locations of springs in the region of Chemech, along the Bol'shaia, the Ozernaia, and the Paudia Rivers. Within each of them steam-emitting fissures can be observed at the bottom of valleys or along low inclines. From them emerge outcroppings of transitional rocks or metamorphic formations of most ancient derivation. And in Kamchatka these formations cover yet glowing volcanic masses in places like independent mountain ranges. The peculiar character of transitional deposits found on Kamchatka (p. 557) was laid down long before the actual volcanic phenomena took place on the peninsula, without direct connection with the latter.

Our comparative overview, above, of the distribution of mountain ranges, the formative relationships of both continents, and the configuration of the islands located in between, revealed that, of the elevation directions, three are predominant. The SW to NE sweep of the Aldanian and the Kamchadal ranges meets us, in a subordinate form, in the western mountains of Bering Island, and predominantly in the lengthwise extension of the Fox Islands. It changes to WSW to ENE, to W to E. It marks out the distribution of the Aleutian Islands and sweeps by way of Alaska to Chugatsk [Kenai Peninsula?]. The American main direction, SSE to NNW to SE to NW, is manifest, other than in the continental mass itself, only in Amchitka and a few other small Aleutian islands, but then characteristically on Bering and Copper Islands. And the latter stand in a more exact relationship to subordinate mountain ranges on Kamchatka, which extend in the same direction.

The different sweep of the coasts and mountains of the two continents with their respective peninsulas could allow us, a priori, to assume a difference in the rock formations of the same. This assumption is furthermore confirmed in a compilation of the sparse geologic data on the high, narrow coastal range of NW America, together with the better known, wide, low Aldan Range, as well as by a comparison of the mountains and volcanic ranges which sweep through and compose the crystalline slate (phyllite), gneiss, and granite which seem to have wider distribution along the American coast, between 50° and 60° N. Lat., than along the Asian coast. And of the type of Jurassic formations found along the Aldan River, we have received none at all from America; only weak traces of the lime-coal found at the coast of Okhotsk. The Tertiary formation, however, which heretofore had been missed on Asia's coast, outside of Kamchatka, seems to be present at Crillon (Lituya Bay), and in the brown coals of Sitka.

Sitka consists predominantly of graywacke. The emergence of this island was caused by granite, syenite, diorite, and porphyry. But until now only traces of the Silurian and Tertiary formations have been found. But such islands are not yet known along the coast of NE Asia, just as little as those that, like Kruzof Island, consist largely of basalt, through which emerged the andesites or trachy-dolerites, and the corresponding types of lava by way of the Edgecumbe Volcano. Only in the northern, nonvolcanic part of the island do the graywacke and porphyries occur.

Alaska [by Alaska, the author means the Alaska Peninsula only] has gradually become a peninsula, formed out of individual islands, or rather it was transformed into such. Along its southeast side there is another, analogous and parallel line of elevations with Kodiak, Afognak, Shuyak, the Barren Islands, and Chugatsk [Kenai Peninsula]. In this respect, too, Alaska differs from Kamchatka, which is a cohesive mass and does not consist of individual islands.

From the estuary of the Sushitnan [Susitna] on the west coast of Cook's Inlet downward, and along the entire SE coast of Alaska, we see the predominant elevations and volcanos pushed hard against the steep and rugged coast. Kalgin Island, opposite or practically at the foot of the volcanic High Mountain [Redoubt Volcano?], is composed of primary shales and eruptive rock. But we sadly lack, until Alaska, any further data and specimen for geologic occurrences. And only by the volcanicity of the High Mountain and of the Ilaman (Ilamna) [Iliamna] Volcano can we deduce the predominance of volcanic types of rock. We can furthermore point out that on the western limit of Cook's Inlet, as well as on its eastern shore, deposits of Tertiary layers are found.

But from the sparse geologic material available to us there seems to emerge the fact, although not with absolute assurance, that on the eastern part of the Alaska Peninsula, especially 30 to 50 nautical miles from the northern coast in the peninsula's interior, from the Sulima [Ugashik] to Naknek and to Lake Iliamna, there is a predominance of granite, gneiss, and primitive shale, then followed by eruptive types of rock (porphyry and diorite). From them arose volcanic kinds of rock, a little farther toward the south coast. Along their southern slope, i.e., on the SE coast of Alaska, mainly graywacke-shale and metamorphic, less eruptive rock is laid bare. On the Four-Summit Mountain [Fourpeaked Mountain], on Alai and Chiginagak Mountains, the volcanic types of rock seem to be more developed, while at the coast between these mountains the conditions might be analogous to Kodiak. That is to say, predominant are graywacke clay-slate, sandstones, and Tertiary deposits. Yet volcanic products might also occur, albeit subordinately. The occurrence of Jurassic layers on the coast of Katmai, and the

Tertiary layers in the interior which cover them, must yet be more thoroughly investigated.

The rest of Alaska, west of the *perenosy* [portages] between Heyden Bay [Port Heiden] and Kishulik [Kujulik] and Chignik Bays, has a predominantly volcanic character. The Veniaminof Volcano stands more in the middle of the peninsula, while the Pavlovskii and Morshovsky [Morzhovoi] Volcanos lie very close to the rugged, highly elevated, island-rich southern coast. The farther southwesterly the progress, the more widely distributed are the basalt rocks. The islands Peregrebnoi [Wosnesenski], Pavlof, and Amak are composed of them where they are supposed to be found even at the foot of volcanos. But this wants substantiation as much as the claim that porphyry and granite occur on the northern slopes of the volcanos mentioned. There is, however, no doubt at all that in many locations, especially along the south coast of Alaska, Tertiary strata, and the brown coal which occurs with them, sometimes overlie volcanic types of rock; and sometimes they are found in clay of volcanic tuff. Where they are absent, diluvial and alluvial formations surround the outer margin of the island-like parts of Alaska. Along the low northern coast of Alaska they seem to be more developed. The distributions of rock types shown on our Map 5 must often be inexact in their detail, because they are hypothetical. On the other hand, it will by and large prove true that there happened at first a granite elevation, which was followed by an elevation of porphyry, and thereupon one of basalt, and finally of trachyte and andesite.

The geologic conditions of Kruzof or Edgecumbe Island seem to recur on the basaltic islands of this region. But Unga's rock types indicate a relationship with both Kruzof Island and Unalaska. On Unga there occur gneiss, clay-slate, itacolumite, diorite, aphanite, diorite-porphyry, diorite-amygdaloid, and Tertiary deposits, which are not absent from Unalaska, either. But basalt, which was shown to exist on Unalaska, we did not receive from Unga.

Therefore an analogy of the geologic composition of Alaska and Kamchatka is for the time being not possible. Basalt rocks are to date completely absent from Kamchatka. Granite-type rocks and primary shales are rarely reported from there. And whether the melaphyres of Kamchatka and Alaska have really the same large distribution, we do not know. Similarly, the chalk formations found in Kamchatka are nowhere indicated either on the NW coast of America, or on Alaska, or the Aleutian Islands. Their presence in Kamchatka (cf. above) was, to be sure, not sufficiently proven either. The Tertiary formations of all of these regions, however, can well be the same. All the fossils which we have obtained from Kamchatka, the Aleutian Islands, the Pribilof Islands, Alaska, Unga, and Kodiak, support this. And Erman's *Tellina dilatata*, and the genera *Crassetella*, *Venus*, *Nucula*, and *Buccinum*, are not absent from several of the localities just mentioned, except for Kamchatka. The Tertiary layers of Kamchatka, however, would in this case not belong to the older, but to the most recent Tertiary epoch. The few Jurassic fossils found in Alaska resemble many of those found along the Lena River and in Eastern Siberia on the whole.

The Kuril Islands adjoining Kamchatka are still too little known for us to compare them in a more detailed fashion with those islands adjoining Alaska. As we go over the islands between Asia and America from west to east, then Copper Island and Little Alaid (Semichi) provide the first specimen or indications of accumulations of sulfur, volcanic sand, and lapilli with augite and olivine. But the same volcanic products are most likely present on Bering Island and Attu as well, since they are present on nearly all of the islands of this chain. It might be presumed that these products of volcanic action could have been washed up to the coast from the bottom of the sea, wherever volcanic eruptions have not been known to have happened within historic times, i.e., the past 100 years. But the earthquakes, which continuously afflict the Commander Islands from Steller's time until the present, and the mountains of Copper Island, which Krenitsyn believed to be volcanos, as well as the peculiar conical shapes of Bering Island and Little Alaid, lead us to believe that eruptive activity, if only volcanic eruptions, happened also on the Commander Islands, just as with the Near Islands, west from Little Sitkin. The first true volcano islands to our knowledge. Albite-lava and lava-bombs we received, probably by mere chance, first from Atka. Obsidian and pumice stone from Siguam [Seguam] indicate trachyte and andesite deposits. Among the rock samples Mr. Voznesenskii sent in, is that trachydoleritic rock which Erman believes to be andesite, although its composition does not indicate this. It is the same rock which is also found on the Commander Islands (Bering and Copper Islands), on the Rat Islands (Amchitka), and on the Andreanof Islands (Amlia). True andesite we did not receive. Therefore, we dare not assume the presence of that type of rock merely on the basis of the cones, which extend like cores from the top-most summits of the highest mountains of Bering Island, nor on the basis of the probably analogous shapes of Little Alaid (Semichi) Island, although we find sodium-feldspar (albite) in several rock formations of the island chain between Asia and America. There is no doubt that Steller's description speaks for elevation craters, from which arise volcanic cones. And it is possible that they were active in the past, as we noted with respect to Copper Island and Little Alaid; and nowhere in the region has a structure of ribs or plateaus been observed in the composition of these elevations and mountains.[125] We do not deny the possibility of an analogy between the mountains of the Bering Islands and the Shivelich Volcano. But it does not seem possible to prove with the present Erman hypothesis, according to which, mistakenly, there is not a trace of lava at the Shivelich (cf. note 123). Star-shaped groupings and similar elevations, as at the Shivelich Volcano, we find neither among the Commander, nor the Near Islands, nor among the Fox Islands where, except for Unimak, the absence of valleys is conspicuous. The coves, especially on the northern coast (Veniaminov, I, pp. 6, 14), are mere ravines, 5 versts wide and 12 versts deep at the most, but where the rock types we know do not represent true andesite.[126] The occurrence of that type of rock on the islands between Asia and America, we can therefore, because of limited knowledge at this time, merely assume to be possible. But we cannot ascribe to them the wide distribution which Erman claims for Kamchatka. At the same time, we do not hide the fact that we received only a few specimens from the summits of the mountains. Likewise, it cannot be denied that the volcanos on the western part of the Aleutian Islands, where albite is more noticeable among the rocks, are presently less active than on the eastern part, where porphyries and basalts are more predominant. This does not yet prove that this phenomenon is the result of the overall structure of the islands, rather than constituting the periodic rest of a part of the large volcanic belt that surrounds the Great Ocean.

Andesite, according to Erman, broke through the layers of graywacke-sandstone and clay-slate. But on Copper Island the latter is found together with jasper—a related rock. This would allow us to deduce the presence of eruptive types of rocks. Mr. Erman might have richer collections from these islands available to him than we do. Therefore, we shall, as he does, maintain the predominance of graywacke formations in that place. On Attu and Amchitka we already see, aside from the clay-slate, the emergence of eruptive types of rock (such as aphanite, diorite-porphyries, diorite, serpentine); and on Amchitka also phonolite and trachydolerite. But Atka is the first island where granite, gneiss, as well as graywacke, jasper, albite, and clay-porphyries, conglomerates, basaltic rocks, and volcanic products are known to exist. A characteristic basalt-like albite-porphyry surrounds the base of the volcanos of this island. From them emerge trachydoleritic rocks, and lava related to them, the former of which are widely distributed on Amlia. If we want to apply to the Aleutian Islands the same conceptions Erman uses to describe the composition of Kamchatka's volcanos, then we should find the andesite mountains on those islands surrounded with porphyries. And depending on the state of activity, quiescent or eruptive, the islands should be composed either of andesite or of porphyry. This assumption we can, however, not make at this time. Although some volcanos might consist of andesite, we feel they are formed analogously with the trachyte cones. This seems to emerge from earlier observations (pp. 126–130) and from a comparative examination of the larger, better-known islands: Atka, Unalaska, Umnak, and Unimak.

The shapes of Atka and Unalaska are formed quite analogously; less so the shape of Umnak, which is in some regard more similar to Unimak. The northern ends or halves of the three islands first named form peninsulas, which in earlier times might have been islands, as can be learned from traditions concerning Umnak and Unalaska. The northern peninsula of Atka has on the whole less rugged coasts than that of Unalaska. But the former could well be compared with a part of the latter, namely the Makushin Peninsula between Captain's Bay and Makushin Bay. On the peninsula of Atka, just mentioned, four true volcanic cones are known to us, which all surround a fifth one, the Kliuchevskaia Sopka (not clearly enough indicated on our Map 5). They rise to an elevation of up to 4,850 ft. (Korovinskaia Sopka); their summits are smoking and on their slopes hot springs and mud craters break forth. Therefore, they should not be mistaken for eruption craters, such as surround the Kliuchevskaia Sopka. The albite-containing types of lava at the Sergeyevskaia Sopka, as particularly also the red types of lava and the obsidian, are indicative of true volcanos. But the hearth of volcanic activity seems to have moved so close to the surface here that strong earthquakes are observed in this location, although several craters of these real volcanos provide vents for uninterrupted escape of gases and steam.

Only two burnt-out volcanos are reported to exist on the northern part of Unalaska. Better known is the Makushin Volcano, which smokes constantly. Although this mountain has not belched fire in historic times, and its crater is located on the plateau-shaped bluntness of its cone, we nevertheless want to number it among the true volcanos, particularly because of its connection with St. J. Bogoslof Island. But the rocks east of Illuluk, which stand together in crater form, are either elevation craters or remnants of eruption cones (eruption craters).

It cannot yet be ascertained if one of Atka's volcanos, or which one of the five, provides an outflow for the gaseous fluids. The Conical Sopka, however, seems to be most similar to the Makushin in location and form. On the southern slopes of both volcanos there are hot springs emerging, and their SW sides fall off abruptly, while the smoking Korovinskaia Sopka has two summits and inclines abruptly on its northern end, and gently on its SW side. On the gentler slopes of the first-mentioned volcanos we usually find types of lava, which in Atka show more characteristic feldspar than is the case on Unalaska. Deposits of obsidian, pumice stone, and feldspar-porphyry indicate andesite elevations. Nevertheless, it is also the place where for the first time melaphyres, numerous augite crystals in lapilli, and basaltic, olivine-containing types of rock occur in larger distribution. Layers of Tertiary deposits are found on the west side of the Conical Sopka and on the NNW and eastern slopes of the Makushin Volcano. According to what we presently know about Unalaska, it seems that from SE to NW in the direction toward the Makushin, there first are found clay-slate and metamorphic rocks, then granite, gneiss, and conglomerates, and finally trachytes or andesites and the newest products of the lava volcanos. But eastward and southward from the porphyries of Captain's Bay toward Beaver Bay, there is aphanite, diorite, diorite-porphyries, and amygdaloid, serpentine rock and other volcanic types of rock deposited. These indications, to be sure, are yet to be confirmed by more exact investigations. But what speaks for it, is the fact that in Umnak and Atka, too, granite-like rocks[127] take on analogous positions. The basaltic rocks, which on Semichi (Alaid) Island are indicated by the presence of olivine or augite in the volcanic sand, are not absent from Atka and Umnak, either. There are, moreover, on those islands outcroppings of albite-porphyry with corresponding volcanic types of rock. The analogy of rock formations on Unga and Unalaska has already been mentioned.

The occurrence of primitive types of mountains (here perhaps merely younger eruptive granite), is characteristic for volcanos in rows. But the undisputed existence of basaltic rocks on Unalaska, etc., indicates elevation craters with or without centered cone, or eruption cones and craters. Umnak with its three volcanos, plus St. J. Bogoslof, which is the link between Umnak and the Makushin Volcano on Unalaska, serve perhaps best to explain how a trachyte or andesite cone can arise on part of a long subterranean canal; how by its crater it stood for a time exposed to the atmosphere. Thereafter it closed up again, while in a different location a new cone developed, or an old, plugged exit broke through again. Unfortunately our information is so incomplete, that we cannot even decide where real volcanos are in action, or if we deal only with volcanic eruptions. Unimak seems, indeed, to have three real volcanos (Pogromnoi, Shishaldin, and Khaginak). All other phenomena might simply be subordinate eruptions. But the frequency and strength of the latter seem to prove here, too (cf. Atka), that the constant opening of one or several escape vents for the gasses (as evidenced by the simultaneous smoking of several summit craters in close proximity of one another) does not have the same pacifying effect in row volcanos, as it has in central volcanos. The Islands of the Four Mountains are perhaps only eruption cones. Yunaska, Amukta, and Seguam are volcanos. But any further pursuit of this hypothesis, viz.: which volcanos, and how many per island, are elevation craters, trachyte or andesite cones, or merely deposits, is bootless. And it certainly is impossible to prove with regard to the little islands. There is, for instance, the island Amaknak in Captain's Harbor. On it are reported to exist basalt and olivine-type lava, as well as obsidian and porphyry, while Spirkin [Sedanka] and Unalaska's east coast seem to consist entirely of obsidian.

It can be fairly well supported that on the west to eastward sweeping Aleutian Islands, especially in the bigger and higher masses,[128] there is a larger distribution of granite rock. Likewise, several, especially smaller, islands are of a basaltic character.

The eastern islands rose above the water level before the western islands did. And older, as well as more recent, elevations declined in strength as they progressed from east to west. Or the more recent elevations encountered less resistance. This caused either a larger distribution of trachydolerites and volcanic products, or an absence of the same toward the west.

No granite rises above sea level on Attu and Amchitka, nor on Bering Island or Copper Island, while diorite, serpentine, soapstone, shale, jasper, and clay-slate, eruptive, metamorphosed rock and graywacke are proven to exist there, and trachydolerites are more or less distributed. We found granite, gneiss and syenite for the first time in our collections from Atka. And basaltic rock, except by earlier indications of the same because of the presence of olivine and augite, we find first on the island just mentioned, and then on Unalaska. But the types of clay existing on almost all of the islands, allow us to deduce neither the existence of clay-slate, nor of basalt, so long as we lack more detailed evidence.

The Pavlof Islands and Peregrebnyi [Wosnesenski] Island consist entirely of basalt; and Amak and especially the Pribilof Islands consist predominantly of these types of lava. On St. Paul feldspar lava is predominant; nor is olivine-type lava absent here, which is found more frequently on St. George. Pumice, obsidian, red and black olivine-containing lava are found along the northern coast of Alaska to the Naknek River, and farther northward to the Kuskokwim River. But it seems that the basaltic character found on the Pribilof Islands, as well as St. Mathias [Matthew], St. Michael, and Stuart Islands is predominant. On St. Lawrence Island basalt and basalt-like lava are found on the west side and toward the eastern side. The southern point, however, is supposed to consist of granite and diorite. Volcanic and eruptive types of rock, graywacke-sandstone, and clay-slate, metamorphic types of mountains, and granite types of rock reach far northward along the coast of the mainland of North America, until we observe hard coal deposits, too, at Cape Thompson. A closer review of the geologic conditions of the latter region would only be a repetition of what has been mentioned in Chapter 3. Useful, however, would be a more comprehensive investigation of these formations on the islands between Asia and America, and on the NW coast of the latter, wherever they are found to contain fossils.

Diluvial formations were demonstrated to exist by the mastodon remnants on Unalaska, the Pribilof Islands, in Norton Bay, Kotzebue Sound, and along the coast farther north of the same. They are also indicated at Cook's Inlet and along Alaska's coastlines. But only after more detailed knowledge of the land can the distribution of these formations be determined with certainty.

Most developed seem to be the Tertiary formations with their brown coal deposits. We find them on Amchitka (Kirilof Bay), Atka (Conical Sopka and Sand Bay [Martin Harbor]), Umnak (Tulikskoi), Unalaska (Makushin, Captain, and Mokrovskoi [Pumicestone] Bay), Akun, Tigalda (Naknek, Moller Bay, Morshevaia [Morzhovoi], Pavlof, Perenosyi [Balboa] Bay, and north of Katmai Bay), on Unga (Sakharof Bay), Kodiak (Igatskoi [Ugak] Bay and Uganik Village), on the Chugatsk Peninsula [Kenai Peninsula] (east coast of Cook Inlet), Altua [Lituya] Bay, Sitka, Little and Big Bodega, Fort Ross, on the estuary of the Sacramento, and from San Jose to Monterey. On the Aleutian Islands, the Pribilof Islands, Alaska, Unga, Kodiak, and Big Bodega[129] the fossils belong to one and the same period, although the indications of the finds might not always be correct. The presently living mollusks of the Bering Sea are very close to these fossils (cf. Appendix I). In some places they are perhaps completely identical. There do, however, occur such significant differences, that we were persuaded, on the basis of the present knowledge of the fauna of the sea, to give an account of all the fossils of the most recent Tertiary formations. In this we are furthermore prompted by the surprising resemblance of these fossils to the fossils from the Tertiary basin of Beauport near Quebec at the Gulf of St. Lawrence, which appear to be closely related to the youngest Pliocene strata found in Scotland (Lyell in: *Transactions of the Geological Society*, LV, 1841, Part I, pp. 135–139). Erman's Kamchatkan Tertiary formations, too, will perhaps with closer observation show the same analogy (cf. [Erman?] p. 120). Herein we find yet another proof that the unity of the present distribution of arctic animals has had its equal in earlier epochs. Finally, we must point out the resemblance of these strata to some kinds of Pliocene Tertiary layers on the northern slope of the Caucasus Mountains. Thus here, too (as with the massive rocks), an analogy of the conditions is possible.

The occurrence of chalk formations should be assumed on the basis of the rock strata of some of the islands and Alaska, and of their ample distribution on Kamchatka. Yet we believe that there is more reason to assign some of the mollusks, which resemble those of that era, to the Tertiary age.

Marginal indications of Jurassic strata are found by us only at the Bay of Katmai, on the SE coast of Alaska [Peninsula].

Formations of hard coal are indicated along the Columbia River, on Vancouver Island, and on Unga Island. Along the northern part of the east coast of the Bering Sea, as well as of the Arctic Ocean from Cape Lisburne to Beaufort, its presence is possible. The occurrence and even the larger distribution of the Silurian system is assumed, and not without reason. We, however, have received no proof of this formation, other than one *Catemipora escharoides* from Sitka (probably mere boulders), and traces from the northern coast of the Bering Sea.

Further than this we would not want to advance in our comparative perusal of the geologic composition of the region described by us. But now several attempts have already been made to determine the age of the elevation of the Aleutian Islands and of the continents opposing them. These we cannot well ignore quietly.

Generally it is agreed that a large part of North America has been drained at the end of the transitional period (neptune, clay-slate, graywacke-sandstone, and limestone). This assumption is founded primarily on the geologic conditions of the eastern half of North America. But it seems to apply to the western half as well, if it is postulated that in the latter region there is a fairly large distribution of azoic shales and Silurian formations. In that case the conspicuous dearth of formations younger than the Silurian, but older than the Tertiary, would support the assumption above. Such hypotheses, however, are premature, because the fact that certain formations are not known in scarcely investigated regions is no proof for the nonexistence of the same. Since we deduce the existence of Silurian shale from indications and analogies, we can with more reason assume that the hard coal and Jurassic formations have larger distributions than we can know to date. Thus, drainage could have happened in different locations during any of the latter ages. The highest mountain ranges on earth, such as the Andes, Cordilleras, the Alps, are proven to be the youngest, i.e., those that have arisen most recently. It is because the thicker the crust of the earth, the more difficult is the breakthrough of volcanic products.

The volcanic types of mountains of the Californian and NW American coastal ranges, and the volcanos of the Columbia Range provide us with certain proof that since that (hypothetic) first period of plutonic elevations, vivid changes of the earth's surface have taken place here in most recent times. The California Peninsula, for instance, seems to have arisen during, or even later than, the Tertiary epoch, to judge by the large distribution of trachytes and andesites. And the occurrence of Tertiary layers on several points on the coast of Monterey to Lituya Bay (Port Français) proves that on this part of the west coast of North America an uplift has occurred after the most recent Tertiary period. If we adhere to the assumption of an older uplift, then a depression below sea level must have occurred shortly preceding the Tertiary period (provided a Silurian formation can be found here in the first place, while the intermediary strata until the Tertiary period are missing), whereupon a second uplift must have happened.

From the copper and galena-containing shales or the primitive types of rock on the east coast of Nutka Sound on Quadra [Vancouver Island], there are on the west coast outcroppings of volcanic types of rock. Then northward there follow clay-slate again, and next to this eruptive types of rocks, which surround the rich coal deposits on the northern coast of this island. Charlotte Island seems to consist primarily of clay-slate and sandstone. But on the northern coast of this island volcanic rock can also be found. On Prince of Wales Island seven volcanos are reported to exist, according to old reports. Consequently, there must be the corresponding types of rock distributed in the vicinity of Bucarelli Harbor, at Mt. Calder, and generally throughout the whole island. But clay-slate and granite are not absent, either. It might, however, be too premature at this time to speculate on the age of these elevations here, as well as at the little-known Sea Alps of the NW coast of America from the Stikine to St. Elias, the Yakutat and Truuli [Kenai] Ranges, where the distribution of slate, mica-schist, gneiss, and granite is perhaps more evident. But here as elsewhere, the general character of a mountain range from which row volcanos arise, can be surmised, without St. Elias, Fairweather, and Crillon needing necessarily to be true volcanos. On Sitka and on most of the other neighboring islands along the southern part of this coast, so rich in islands, the granitic syenites, diorites, and porphyries have pushed up predominantly graywacke-shales and sandstone. But on the volcanic island Edgecumbe [Kruzof], there arose from the porphyries basalt, trachydolerites, (andesites) and lava. At the bay near Chugatsk, the volcanic types of rock will surely occur in larger distribution up to the Wrangell Volcano, if Fidalgo's eruption location can be found again.

According to von Buch, the mountain ranges in rows are viewed as masses that have arisen from a large fissure by the action of the black augite-porphyry (melaphyre). And from just such fissures the row volcanos arose above the ridge of the mountain range, or perhaps directly from the interior of primitive types of rocks. In that manner the volcanos located close to the coast: Uyakushatch (the high mountain) [Redoubt Volcano], Ilamna or Ilaman [Iliamna], Pavlovskii, Medvenikovsky, and Morshovsky form the highest points of a narrow mountain or eruption range. The Veniaminof Volcano alone is situated more in the middle of Alaska. In several locations these volcanos and their products are surrounded by basalt, more eruptive rocks, granites, and predominant outcroppings of primitive shales and Tertiary layers. And thus the geologic conditions of the Aleutian Islands seem to be repeated on the island-like partitions of Alaska. Here, however, the volcanic powers were more forceful. They encountered more resistance. And during their constant activity they have brought up lower-lying Jurassic deposits as well. But the widely distributed, often horizontal, Tertiary layers prove that the major uplift of the volcanos of Alaska and of this peninsula itself took place prior to the Tertiary epoch. But along the north coast of the peninsula the rest of our present ocean inhabitants are now laid dry by way of a gradual rising of the land.

According to Erman (III, pp. 315–317):[130]

> Kamchatka, or that part of the earth's surface through which the Shivelich and the Middle Range erupted, have during the first geological .periods already been pushed up by plutonic kinds of mountain masses. Thereafter, however, the sedimentary rocks that formed in later periods failed to cover the original strata in most places. The diorites of this region, their compositon, and their differently composed structure and appearance are completely similar to those, which in most Northern Asian mountain ranges (from the Urals to the Aldan, to the Okhota) have broken through the oldest transition layers (the Silurian or graywacke formations), some of them immediately after their formation.

Some, like those in the Aldan Mountains, have been transformed into rare crystalline rocks, and this already during the time of genesis. The actual granite, on the other hand, seems to have a very subordinate distribution among these oldest plutonic masses of the peninsula. The same is true in the Urals and in the mountains of the eastern part of the continent. These granites, as well as the granite in many other regions of the earth, have not formed one single mountain range. They formed instead the broader, not much elevated plain, on which later rock eruptions arose, as they reveal by their mighty lava streams the continuously active smelting heat that exists in their lower regions. The Shivelich Volcano and those andesite summits which resemble it most, were the first among the volcanic mountains of Kamchatka to be elevated. But this activity still falls into the most recent Tertiary period. After its formation, not a single one of those more general transformations of the earth's surface can have taken place (p. 317). Those masses, however, which consist of labradorite and augite, have emerged from a greater depth, which occurred first in that range of volcanic mountains west of the Shivelich (Middle Mountains), where other than lava, there also are found those crystalline mutations of these fossil rocks. They are related to the trachytes by their light grey color and by the predominance of their feldspar-like ingredients. Thereafter they occur only in those unequally higher and still active cupolas of the eastern range of the peninsula.

Hence, none of the three epochs of verifiable volcanic activities in Kamchatka: that of the andesite eruptions, that of the activity of the Middle Range, and the emergence of the presently active conical mountains of the eastern range, can be assumed to be older than the Tertiary mountain ranges. The ruins of volcanic fossils which are found in the western half of the peninsula, in that group of stratified formations, and even in younger chalk deposits (?), can only be explained by a fourth and earlier class of eruptions. This was probably connected with the formation of the amygdaloid, many of the other and more distant, melted products of which are now long buried beneath the sea.

These options of Mr. Erman's do not necessarily have to be believed, which emerges from the fact that his geologic observations and indications concerning the occurrence and distribution of rock formations want substantiation.

There is perhaps more reason to assume for the region extending from Alaska via the Aleutian Islands and the Komandorskie Islands to Kamchatka, that a range of granite and shale had protruded above sea level during the first geological periods. But too little is known about the relationship of the eruptive rock type to the granites and shale rocks and to the conditions of the deposits of the layered strata, which contain fossils, to make at least fairly accurate determinations concerning their age. We therefore completely refrained from speculating on the latter, except where the question: were the continents of America and Asia once connected? makes it perforce unavoidable.

Since the beginning of the sixteenth century people have been preoccupied with searching for the longed-for, easily navigable connection from the Atlantic to the Great Ocean. First it was the Anian Strait, then the Fuca Strait, etc., until in the eighteenth century this hope was dashed. In the same manner people continued to argue for the existence of a connection of Asia with America,[131] even after the voyages of Bering, Cook, and Vancouver. Franklin's, Ross's, and other voyages left finally no more doubt about the fact that these continents were divided, and about the island formations of America. Thus, the agile human spirit raised a new geographic problem of prehistoric time, after the solution of the former question: Have not the continents perhaps been connected in earlier ages; and when and where had they been connected?

Dr. Stein (*Trudy Mineralogicheskago Obshchestva*, ch. I, p. 387) answers this question decisively in the affirmative. He believes that the Aleutian Islands are remnants of a rock ledge which once surrounded, together with the coasts of Asia and America, an ocean cauldron, within which the islands St. George and others constituted shallow places and sub-oceanic hills and mountains. Volcanic powers and a deluge from the south caused the destruction of that dam and the breakthrough of the waters. To this day, strong currents, low and high tides, storms, and the deteriorating influence of the atmosphere cause a progressive diminishing of the circumference and elevation of the islands.

Father Veniaminov (I, pp. 102–107) believes to be able to assume with certainty that the islands in the region of Unalaska are not of volcanic origin, because on some islands, whole mountains consist of granite. He believes the islands are remnants of the mainland, which extended between Kamchatka, Chukotskii Nos, and America. Or the islands were larger and connected with one another, with a few volcanos on them, which collapsed all at once or at different times (*obrushilis' v bezdnu*). In this manner they produced straits and bays.

> Elevations are observed on the northern coasts of the islands, and also along the entire American coast of the Bering Sea. They extend even as far as the base of the mountains which are located in the interior. It is only necessary to seek out an elevated point, in order to observe from there how a sea with waves has seemingly been arrested, the waves converted into sand and mud, which now is covered with vegetation. In the midst of these petrified waves there are here and there high islands, which seemingly have been lifted from the ground. But they, too, are nothing else but remnants of the mainland, which had been submerged in water, only to re-emerge at this time. All of these uplifts prove that where there is the Bering Sea at this time, there once was dry land, which had

been destroyed and cast into the sea by the power of subterranean processes. Thus the islands within this sea are only remnants of the same. Nor is the emergence of the island St. J. Bogoslof any proof concerning the other islands, inasmuch as it is so small and separate.

Thus, Dr. Stein assumes a rock dam, sweeping along the region where today the Komandorskie and the Aleutian Islands extend, to have connected both continents and at the same time divided the Bering Sea from the Pacific Ocean. Veniaminov, on the other hand, assumes that the entire Bering Sea has once been dry land.

It seems, however, that neither of the two authors was adequately acquainted with the geologic composition of the islands, nor with geology as such, to do much more than raise hypotheses with regard to these regions, relieved as they were of the weight of observed facts or other data. In our opinion a satisfactory solution to this question is not yet possible. Ethnographic studies and research with regard to animals is not far enough advanced to provide us with adequate explanations. The former reach at best as far back as the dawning of humanity, which is our most recent geological period. And even with the most exact knowledge of the presently living fauna and its species, an insight into the mysterious history of the development of the organisms of our earth cannot be accomplished without the study of paleontology. Thus not without reason will the decision concerning this question be primarily left to the geologist. But where the question concerns a former connection of the continents, the geologist cannot consider it apart from observations of certain geological periods. In our case he can arrive only at negative conclusions and relative determinations of time periods, because the geologic understanding of these regions must be called most incomplete. We cannot take conditions of stratification into consideration, but only the presence or absence of formations as such. Whenever water was part of the alteration process of the earth's surface, the moment of the emergence of the land can usually be demonstrated, but not the incident of the sinking of landmass. And finally, it is not possible to invest geological time periods with certain time values, expressed in numbers.

Granite, gneiss, azoic, and perhaps Silurian shales can well have formed a mountain range during one of our earth's oldest epochs, where now we see the Aleutian and the Komandorskie Islands. It could well have extended above sea level, connecting Asia with America, because along the entire inner rim of the Bering basin we have on the slopes of the mountains, etc., no proof for the deposit of formations from the periods between the Silurian and the Tertiary ages. But this lack provides us at the same time with proof that granite and shales, assuming that they have gradually risen above sea level after the latest earth revolution, must have been clear above sea level well before that revolution. Granite does not seem to rise to significant elevations. Shales are more likely to do that.

North of the Bering Strait (Cape Lisburne to Cape Beaufort), and on the southern side of the Alaska Range (Unga, Chugatsk?) we find the hard coal formations. During that period, therefore, these places were part of the outermost rim of the range of the Bering valley. They were at that time less elevated above sea level than they are now. The valley was covered with lush growth of ferns and shave grass, where today we look in vain for vegetation the size of trees. We cannot yet decide if eruptive or younger rocks lifted that range above sea level.

Jurassic strata along the north coast of the Pacific Ocean or the south coast of Alaska (Katmai), and in Asia have at first become known at the western slope of the Aldan Mountains (Lena). But we know extremely little about their stratigraphy and about their relationship to the eruptive types of rocks. Therefore, we can merely show that the breakthrough of the porphyries and a corresponding elevation of Alaska, the Aleutian Islands, Kamchatka, etc., could well reach into this period. Should Jurassic formations eventually be found along the inner rim of the Bering basin as well, then this would indicate that a change of the Aleutian rock dam, or granite and shale mountain range could already have happened during that [Jurassic] period. Should Jurassic and chalk strata prove to be absent, then the Bering Sea could well have been a drained valley basin during these geological epochs. It would have been surrounded by not very high mountains, with rivers draining through them into the Pacific Ocean.

But after the chalk period [Cretaceous], conditions must undoubtedly have emerged which provided for a large expansion of the Tertiary sea within the Bering valley. The occurrence of the very same Tertiary fossils on the Aleutian and Pribilof Islands, and perhaps on both coasts of Kamchatka, etc., proves that the causes for that breakthrough or for the development of the Tertiary floods must be searched for during or preceding this period. Basaltic eruptions might at first have happened close to America (Alaska). Uplifts of andesite and trachyte must have come later. But this did not happen until after the Tertiary creation had experienced in part a violent demise, or it had outlived itself. Mighty mastodons lived then in the north. Only then did the volcanos and volcanic phenomena, which are active to the present time, while at their beginning enacting a much more grandiose and horrible influence, bring about the latest flood (diluvium). This flood caused the demise of those mastodons, as it deposited their remains along the entire rim of the Bering basin, where they sank into the ground.

Since the Tertiary epoch, a connection of Asia with America could only be imagined in that place where presently the continents lie closest together [now thought of as the Bering Land Bridge]. That is between the Asian East Cape and the Nykhta Promontory (Wales). No traveler since Cook (III, German I, p. 145) has missed noticing the similarity of the two coasts of this region. Kotzebue (I, p. 156) mentions on the occasion of seeing the East Cape: "The terribly ruined rocks remind a man of the revolution of the earth which once occurred here. Both appearance and situation of the coast render it probable that Asia and America were once connected. And the Gvozdev [Diomede] Islands are the remnants of the connection between the eastern cape and Prince de Galles." The geologic composition of the St. Diomede Islands (Gvozdev) and of the opposite coasts is, however, not known. Therefore, we cannot determine if the division of the continents happened simultaneously, earlier, or later than the invasion of the waters from the south. Should the indicated islands be composed of volcanic types of rocks, which is more likely judging by the conditions around Asia's St. Lawrence Bay, then it could be assumed that after the Tertiary epoch the eruptions had caused a rupture of the rock dam and a breakthrough of the waters. These problems might perhaps be solved by a close investigation of the geologic composition of these coasts, whereby shore ledges, abrasions, and cuts will have to be considered, as well as the ocean currents. According to the latest information we know only that the surface current, 12 ft. deep, flows from the Bering Sea into the Arctic Sea; and a counter-current is supposed to flow along the bottom.

Let us attempt to unify what we have said on these pages into an, albeit hypothetical, picture. During the oldest epochs a mountain range consisting of granite and shale might have surrounded the Bering basin more distinctly cohesively than it does now. Presently we have learned of clay-slates, steeply uplifted, probably at a later time, along the northern part of the east coast of that basin to Norton Sound. On the Kuskokwim River there is granite. But then from Alaska through the Aleutian Islands and the Komandorskie Islands to the sources of the Yelovka River (Kamchatka), and in the Aldan and Chagaktakh Mountains, granite or primitive shales have been found. During the hard coal period the continents were perhaps still connected at the present Bering Gate. But the breakthrough of the diorites and porphyries has happened early, and perhaps more violently on Kamchatka than in Alaska. The Komandorskie and Aleutian Islands correspond to that breakthrough in diminishing progression south of Kamchatka and west of Alaska. After a longer dormancy there followed then, in or prior to the Tertiary period, depending on a yet to be investigated stratification of Tertiary lodes, basaltic eruptions, especially in the region of Alaska, in the middle of the Bering basin (Pribilofs, St. Matthew, [St.] Lawrence Islands), on the east side of the same (St. Michael, Stuart, Norton Sound, Kvikhpak), and perhaps on the northern side, too (Cape Nykhta and East Cape). With them arose, not much later, but in larger dimensions, the Aleutian Islands on one hand, and the trachyan Middle Range of Kamchatka on the other. Finally, i.e., prior to the diluvium, the Kurilian and Aleutian row of volcanos broke through the clearly outlined fissures in the porphyry and trachyte uplifts of the Kamchatkan and Aleutian Mountains. These gave to the islands their essential form, which, by virtue of their continuing activity, has not changed significantly up to the present time.

Within historic times, that is, within the span of 100 years, only one or two small islands (St. J. Bogoslof and Amnak) emerged within the chain of the Aleutian Islands. They did not cause any significant alterations in and on the wide sea basin. If we would want to apply a similar phenomenon to the rest of the Aleutian Islands, which obviously form an integral system,[132] then we would do the wrong thing. First of all, it can be noticed that the volcanic forces are diminishing here. Therefore, island formations were more frequent in the beginning of their activity. At that time, too, several islands must have arisen simultaneously and in shorter intervals. But now every formation of a new island within the chain of volcanos from on top of the indicated fissure which runs the length of them, must be called a chance occurrence, happening ever more seldom, because every eruption on an already existing island through an already existing opening to the atmosphere of the volcanic hearth is the same phenomenon. And if furthermore we notice an enlargement of the landmass in the eastern part of the Aleutian Islands and along the east coast of Bering Sea, we notice by contrast, that toward the Asian coast with Bering Island a significant diminution of the land is observed.

It is well known that the latest powerful events and alterations of the earth's surface, which include the elevation of the Aleutian Islands and the accompanying diluvium, are dated prior to the advent of the human being. Therefore, since the time man appeared in the northern regions, he found the distribution of land and water little different from the way in which it presently appears. It seems as if a population movement occurred from Asia (Japan and the Kuril Islands), the so-called cradle of mankind, in an eastern direction along the Aleutian Islands.[133] From America there was a westward trek by the Chukchi across the Bering Strait. The Bering Gate proved such a small obstacle to communication between the two continents, since the time man first approached it, that the close relationship of the people of this region as well as that of the animals seems easy to explain. About the early history of the Japanese, particularly of their nautical and trade endeavors, we know too little. Thus, we are not forced to dismiss the possibility of their voluntary or chance migration to the Aleutian Islands. These indications alone make it plain how little we can expect ethnographic or zoological studies to contribute toward a solution of the question whether America and Asia have once been connected. Nevertheless, it will be of interest to many a reader to learn something about the populations of these regions from the unpublished research reported by Mr. L. Radloff,[134] and furthermore to become acquainted with the state of our past and present knowledge concerning the distribution of animals in this part of the inhabited north.

The numerous languages of the natives of Russian America seem, so far as can be judged from the limited resources available, to belong to a few larger language and ethnic families or groups, who all are more or less closely related with each other. Almost all the coastal inhabitants of the Arctic Sea, so far as it is known along America's coastlands, as well as those on the northeast and west coast of the continent to the mouth of the St. Lawrence River in one direction, and to the proximity of the Atna or Copper River in the other, belong to the widely distributed Polar People, called Eskimos. And their languages differ only in dialects. Across the Bering Strait the settled Chukchi apparently represent an Eskimo branch, judging by their language. It was the relationship of the languages and the similarity of occupation, way of life, and clothing, which can hardly be much different among these fisher-peoples of the Arctic, as well as the fact that the Chukchi have become known earlier than the others; all of which gave rise to the frequent controversy whether Asia or America was the motherland of this ethnic group. For America speaks the much larger distribution of the same on that continent, and the analogy of their language with the rest of the American polysynthetic languages, as well as the tradition current among them, of an immigration from the east, and finally the circumstance that the vocality of the Chukchi language is very close to the Eskimo dialects in America, while among the languages of their mighty neighbors in Asia no relationship of that kind can be detected. Along the northwest coast of America, on a mostly narrow margin of coastland, the Eskimos are known by different names: Maleigmiut, Chnagmiut, Agulmiut, Aglegmiut, Ugashenzie (Severnovskie), Kadiakians, Chugachie, Ugalakhmiut, and others.

From about 195° to 169° W. Long. (Greenwich), the Pribilof, Fox, and Andreanof Islands, as well as the western tip of Alaska with the Shumagin group, are inhabited by Aleuts. So far as their language is concerned, they seem to be as different from the Eskimo tribes as they are similar with respect to their way of life. Hence, for instance, the noteworthy transfer of their name to the Kodiak Islanders. The structure of their language has much in common with the character of the American languages. Nevertheless, in contradistinction to the closely related Eskimo dialects, only a very small number of words which sound alike can be found in these two languages. According to their tradition, they have immigrated from the west. And this legend is almost unanimously verified by all observers. They discern in the body stature of the Aleuts decisively East Asian traits; and their facial features remind them distinctly of those of the Japanese.

The interior of the continent, which corresponds to the coastlands of Russian America, is occupied, so far as it is known, by many small bands, who all seem to belong to the tribe Ttynai (Kenaians). This tribe reveals genuine American types in outer structure as well as in language, so far as it can be determined. At Kotzebue Sound, Norton Sound, and especially along the bay that bears their name, Kenai Bay (Cook's Inlet), they live along the coast and are distributed over an area that sweeps from NW toward SE, where they are known by many different names. Most of them live on the upper run of the Kvikhpak and Kuskokwim Rivers, as well as along the watershed of the latter and the tributaries of the Nushagak.

So far as the zoological investigations are concerned, Buffon [? name illegible in available copy] already deduced an earlier connection of the two continents from the similarity of the animals on both of them. Nevertheless, we have shown that a communication of the presently living fauna was possible even without the continents being connected. Therefore, from among the recent classifications, we emphasize only those mammals which are probably identical and therefore a part of this question (cf. Brandt in Chikhachev's Altaian Journey, Sec. IV, p. 449).

Ferae: *Ursus arctos* s.u. *ferox?* (Middendorff in *Bulletin phys. mathem. de l'Acad. des Sc. de St. Petersbourg*, Vol. VIII, p. 229); *Canis lupus*; *L. lagopus*; *Felis lynx?* (Schrenk: *The Lynx Types of the North*, Dorpat, 1849); *Putorius erminia?*

Glires: *Lemnus lenesis* s. *torquatus* s. *hudsonicus*; *Castor fiber*.

Ruminantia: *Cervus alces*; *Ovis nivicola seu montana?* (Kamchatka and Rocky Mountains); *Bos urus* s. *bison?*

After this latter digression, for which we apologize to the reader of geology, we close our work, which had grown to a larger volume than we had originally intended. We hope to have brought about an appreciation of the rewarding aspects, as well as of the difficulty of a more exact investigation of the regions here described. And we hope to have motivated a few natural scientists, inspired with a passion for their science, to undertake a journey to those regions—a journey dedicated soley to scientific pursuits. The rare visit of points far removed from each other on the part of occasional merchant ships is not enough. Anybody who would attempt to promote scientific understanding of these regions in any essential way by these means, would find out too late, that it is a frustrating endeavor.

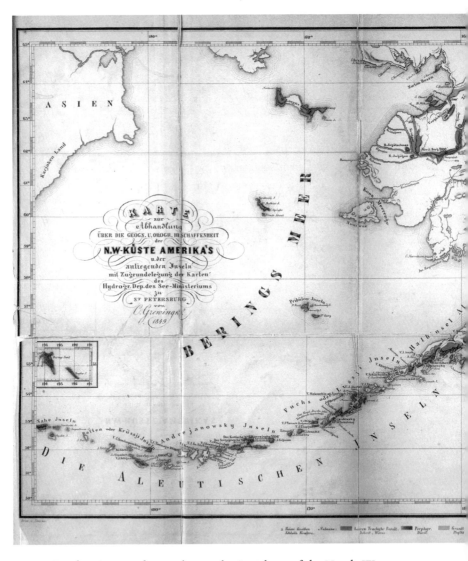

Map 5. For the essay on the geodesy and mineralogy of the North-West Coast of America and nearby islands, based upon Admiralty Hydrographic Department maps in St. Petersburg. C. Grewingk, 1849. [Original in color]

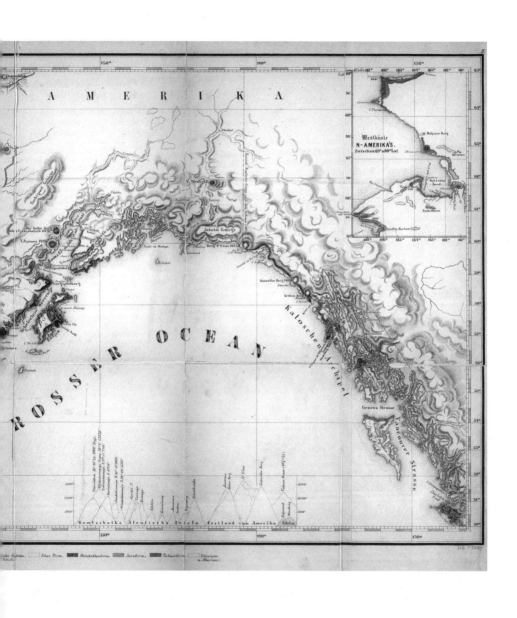

AMERIKA

ROSSER OCEAN

Kaloschen Archipel

Vancouver Strasse

Genua Strasse

Berg St. Elias 16758

Jakutat Gebirge

Westküste
N-AMERIKA'S.
Zwischen 65° u 60° Lat.

Kamtschatka Aleutische Inseln Festland von Amerika Sitcha

APPENDIX I

→> <←

The Fossil Remains of Minerals and Plants Found to Date Along the West Coast of North America and the Aleutian Islands

Bituminous Coal Formations

Entrochite or encrinite limestone with *Lithostrotion* or *Madrepora basalti-formis*, *Flustra*, *Tubipora*, *Productus martini*, *Dentalium*, trilobites from Cape Thompson to Buckland, cf. pp. 48–49.

Mountain lime (black anthraconite, and clay-slate) with *Tubipora*, *Madrepora*, *Terebratula*, encrinites from Cape Lisburne to Buckland, cf. p. 49. From the same location, according to the specimens sent by Fischer and Kupreianov; cf. p. 49: *Cyathophyllum flexuosum* Goldf., *Cyath. caespitosum* Goldf.

Turbinola mitrata His. (Lethea Suecica, p. 100, Tab. XXVIII, fig. 9; syn. *Cyath. ceratites* Goldf. Petref., p. 57, Tab. XV, fig. 12; and *Turbinola striata* d'Orb, *Voy. dans l'Amerique Meridional*, Tab. III Geologie, Pl. 6, fig. 4 and 5).

Caryophyllia truncata His. (l.c., p. 101, Tab. XXVII; syn. *Cyath. dianthus* Goldf. Petref., p. 54, Tab. XV, fig. 13, Tab. XVI, fig. 1; Murchison *Silur. System*, II, p. 691).

Sarcinula sp. (?), *Cyathocrinites* (?), Brachiopoda, whether *Spirifer*, *Orthis* or *Terebratula*, not more precisely determinable.

These definitions might not have been completely exact because of the incompleteness of the specimen. Nevertheless, the polyps indicate undoubtedly the presence of the Silurian formations as well. We, however, follow Buckland's definition of the mountain-lime formations, because his specimens were not available to us, and because the deposits of bituminous coal at Cape Lisburne cannot be exactly verified.

The *Catenipora ascharides* Goldf. from Sitka (re p. 93, which indicates Silurian strata, is contained in a detritus.

Jurassic Formations

Ammonites. Fam. *Macrocephali* Buch.

The volutions expand quickly, especially toward the margin, and they enclose themselves strongly; therefore, only a small, deep umbilicus; unwedged, broad back. The underside has almost as many lobes as the outer side. And the lowest side-lobe always extends over the side edge.

Ammonites wosnessenskii (nob.) Plate IV, fig. 1, a–d.

Volutions almost enclosing; the sidewall angle sinks almost vertically into the inner windings. The umbilicus is therefore deep, formed like a top; and the convolutions extend only a little way.

Measurements:
Height 1–2.5 inches
Growth in convolution 12m/8m = 1.5
Growth in width 28m/11m = 2.5
Growth in disc 50m/24m = 2.08
Thickness 24m/28m = 0.85
Navel width : breadth at the mouth = 2 : 3

Twenty-eight to thirty ribs of one volution begin weakly at the umbilicus. They become sharper at the side edges. After one-fifth of the distance from the outer margin of the umbilicus to the back, usually two, but sometimes only one side-rib is taken up. They are all uniform, do not cut in deeply, and thus extend across the round back, forming a barely noticeable forward-drawn arc across the same.

Lobes and saddles. Plate IV, fig. 1, c and d.

The dorsal lobe is only a little deeper than the upper lateral (L): the same relation between L and l. Two auxiliary lobes on this side of the suture line (NL). Thus the auxiliary lobe, a, surrounds the outer umbilicus rim (NR). And the second, indistinct auxiliary saddle, α', surrounds the inner margin of the umbilicus, or the suture line. On the tripartitioned dorsal saddle (DS) the side lobes are developed in such a way that they fall into one line.

This ammonite is similar to *A. polyptychus* (Keyserling's *Petchora Reise*, p. 328, Tab. 21, fig. 1–3 and Tab. 22, fig. 9) from the Jurassic deposit at the Olenek River. The back, however, is not as flat, and the tubercles of the ribs are missing. This, therefore, is a type between *Am. pol.* and *A. macrocephalus* (Hervey, s. *tumidus*, cf. Quenstadt, *Petr. Deutchlands*, p. 183) found in brown Jurassic beds. Its growth of disc equals 2, and thickness equals 1.

Plate IV

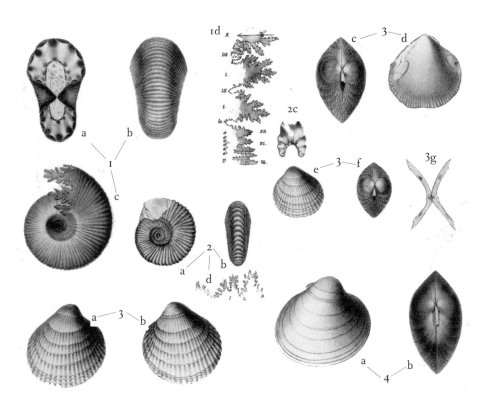

Fig. 1 a–d *Ammonites wosnessenskii*
Fig. 2 a–d *Am. biplex*
Fig. 3 a–g *Cardium decoratum*
Fig. 4 a, b *Card. greenlandicum*

Place of discovery: south coast of Alaska, at Katmai Bay, cf. p. 26. Chamber walls and bowl consist of brown iron ore, which has developed from deteriorating pyrite. Little remains of the epidermis, glistening. Chambers filled with white calcspar.

Ammonites. Fam. *Planulati* Buch.

Back and sides rounded out and not wedged. Volutions little enclosed and mostly pressed together. Radial folds in the middle of its height: one or two, and often bipartite. Without keel over the back, running together from both sides; without knots at the division centers. The upper lateral lobes take on an oblique backward direction.

Ammonites biplex (?) Sow. Zieten. Plate IV, fig. 2, a–d.

Pressed-together shape. Convolutions two-thirds, enclosing. Navel flat. Collar-shaped cuff and ear is indicated.

Measurements:
Size 0.75–1.00 inch
Increase of discs 32m/15m = 2.13
Thickness 15m/12m = 1.23 [*sic*]
About thirty sharp ribs begin from the outer margin of the umbilicus. They dichotomize midway between the distance of the same and the back, and they form on the latter a deep, protruding arc. In different specimens they stand sometimes closer together, and sometimes more apart. In the latter case they are thick and rough; lobes and saddles have sharp points. The upper lateral lobe (*L*) is the deepest. Suture lobes are long and bent. Auxiliary lobes, auxiliary saddles, and ventral lobes on the ventral side correspond to the major lobes and saddles on the dorsal side.

This Jurassic ammonite is also similar to *A. jeanotti* (d'Orb, Terr. Cretaces, Pl. 56) or *A. involutus* of the white Jura (Quenstadt, *Petr. Deutchlands*, p. 165, Tab. 12, fig. 9). But the ribs of our specimen are simpler, cut in more deeply, and on the back more pulled forward. The disc-increase of *A. involutus* is 2.26 and the thickness 1.70.

The place of discovery is the same (cf. p. 26). It is completely transformed into brown iron. Meyen (Nov. act. phys., tom. 17, Tab. 47, fig. 1 and 2) brought an *A. biplex* from the base of the Maipu Volcano in the Andes, southeasterly from Valparaiso. It was the first indication of the existence of a Jurassic formation in America.

Together with *A. biplex* we received (cf. p. 26) belemnite fragments: *B. paxillosus* (?). They, as well as the *A. wosnessenskii*, leave hardly a doubt that Jurassic formations are extensively in existence in the western half of North America as well.

Tertiary Formations

Cardium L.

Cardium decoratum (nob.) Plate IV, fig. 3, a–g.

Testa suborbiculari, radiatim costata, costis 28–30 rotundato applanatis, glabris, interstitiis planis, latitudine costas fere aequantibus inter se separatis, transversim concentrice 7 incrementi periodis regularibus (lege symmetriae distantibus) interruptis, intus striato; margine crenulato, cardinis dentibus obsoletis.

Shell: even-valved, circle-shaped to obliquely egg-shaped, hump toward one side (one-seventh of half the width) moved forward, protruding little and a little rolled in. Maximum thickness a little above the middle. Lunule heart-shaped, ligament on the outside, hinge teeth indistinct, and usually only a thickened tuberosity noticeable (Plate IV, fig. 3g). Twenty-eight to thirty elevated, smooth, flat (not rounded), radial ribs are a little narrower at the hump than the equally smooth intermediary grooves. But downward they grow wider, until at the indented margin they have the width of the intermediary grooves (1.75 m. on old individuals). The lines of growth occur symetrically on specimens of different ages. Toward the rim they stand more densely together, and after the seventh line of growth they follow each other almost immediately. They describe wave-shaped lines, or (in old individuals) angled lines, inasmuch as they progress upward on top of the ribs, and downward in the grooves. Shell seldom over 1 m. thick; inside clearly striped lengthwise, and with deep grooves at the rim. Muscular impression and mantle scar indistinct (Plate IV, fig. 3d).

Measurements

Length	Width	Thickness	
45 m.	48 m.	31 m.	= 1 : 1.06 : 0.66
25 m.	27 m.	18 m.	= 1 : 1.08 : 0.72

Compared with fossils in Europe, this cardium is most similar to *C. pallassianum* (Basterot, Bassin tert. du SE de la France, p. 83, Pl. VI, fig. 2), *C. multicostatum* (Brocchi, p. 506, Pl. XIII, fig. 2, and Basterot, Bass. tert., p. 83, Pl. VI, fig. 9), but especially the cardia from Uddenwalla in Sweden. Among the American ones, the *C. auca* from the Tertiary strata of Chile are lacking lines of growth (d'Orb, *Voy. dans l'Am. Merid.*, T. III, Pl. 13, fig. 14, 15). It seems that those cardia, both living and fossilized, which are found in St. Lawrence Bay and at Beauport near Quebec, are most similar to this one (Transact. of the Geol. Soc. VI, 1841, Part I, pp. 135–139). Among other living cardia, the one most similar to *C. decoratum* in valve construction and measurements[135] is the *C. groenlandicum* (cf. Middendorff's *Malacoʒoologia*

Rossica, p. 557, Tab. XVI, fig. 6–9). Ligament and sculpture resemble the *C. californiense* (l.c., p. 556, Tab. XVI, fig. 23–25). It is found in the Bering Sea, near Unalaska, and Sitka. The latter, however, has forty-six to forty-eight convex ribs.

Occurrences: on Unga (Sakharof [Zachary] Bay, p. 54), Kodiak (Igatskoi Bay, Cape Tonkii [Ugak Bay, Narrow Cape], p. 52), Alaska (Pavlof Village at the bay of the same name, p. 28), Atka (p. 87), St. Paul (p. 109); in volcanic (?) tuff.

Cardium groenlandicum Chemn. Plate IV, fig. 4, a, b.

Testa tenui, laevi, sulcis linearibus radiatis, subdistantibus, magis minusve, imprimis media in valva obsoletis, posticis solis distinctis, cardinis dentibus omnibus magis minusve obsoletis.

Broader and less thick than *C. decoratum*. Muscular impressions and ligament the same. Hinge indistinct. Longitudinal striping very weak, a little clearer at the lunule. The latter more flat and less heart-shaped than in *C. decoratum*. Lines of growth indistinct.

Dimensions:

Length	Width	Thickness	
45 m.	53 m.	30 m.	= 1 : 1.17 : 0.66
44 m.	50 m.	29 m.	= 1: 1.13 : 0.65
43 m.	50 m.	28 m.	= 1 : 1.16 : 0.67
		Average	= 1 : 1.15 : 0.66

Occurrence: together with *C. decoratum* on St. Paul (p. 109), Unga (p. 54), and Kodiak (p. 52).

Cardium aleuticum Girard. Plate V, fig. L, a, b.

(According to Erman's *Archive*, Vol. III, 1843, p. 546, Fig. 9 a, b. This description was given in full, together with the illustrations, because this species was not available to us either from Atka or from any of the other Aleutian Islands.)

Shell almost circular, hump protrudes only marginally, not rolled in, and therefore flat. It stands a little obliquely; therefore the shells at the lunules are a little flatter, while at the front rim they fall off a little more steeply. About twenty ribs extend from the hump across the shell. They are covered with fine lines of growth, which, however, form strong wrinkles in three or four places on the shell. The longitudinal ribs are very weak at the hump. Only toward the middle do they protrude so far that they are just as high as the ribs are wide. On their dorsal sides the ribs are rounded off, and the grooves are only half as wide as the ribs. The shell is very thick, at least 0.5 inch on the specimen of 1.5 inches in size. The inside is smooth, and only the rim is toothed. The frontal muscular depression is inwardly not

Plate V

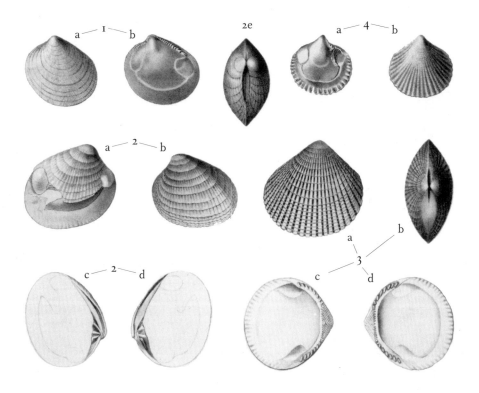

Fig. 1 a, b *Cardium aleuticum*
Fig. 2 a–e *Venerupis petitii* var.
Fig. 3 a–d *Pectunculus kaschewarowi*
Fig. 4 a, b *Nucula ermani*

rounded; it forms instead a point toward the hump. And behind this lies a row of small muscle-depressions all the way into the hump.

Occurrence: in coarse volcanic tuff on Atka (p. 87).

Comment: it is quite possible that this cardium belongs in part to the *Venerupis petitii*, the description of which follows hereafter, and in part to *Pectunculus kaschewarowi*, because the hinging apparatus is not mentioned in Girard's description.

Venerupis Lamk.

V. petitii Desh. var. Plate V, fig. 2, a–e (cf. Middendorff, *Malacozoologia Ross.*, p. 567, Tab. XVIII, fig. 11–13).

Testa transversa, ovato globosa, cordiformi, radiatim tenuiter costata, concentrice sublamellosa (?); cardine incrassato, dente mediano, interdum etiam postico subbifidis; margine subcrenato.

Shell even-valved, obliquely egg-rounded. Hump a little more to the side (three-sevenths of half the width) than in *V. petitii*; and it protrudes a little farther, not involute, and layered together. Ligament stronger and longer than in *V. pet.* Posterior muscular depression with shallow, pointed bay, hinge teeth 3/5.

Dimensions:

Length	Width	Thickness
42 m.	45 m.	29 m.
38 m.	44 m.	27 m.

[Average] = 1 : 1.10 : 0.7

Shell lamellose, to 4 m. thick, covered with numerous narrow, densely standing radial ribs, which are never wider than the grooves in between. The lines of growth are numerous and close together, not high, but clearly grooved. Inside smooth, and grooved only along the lower margin.

Occurrence: Alaska (p. 28), Unga (p. 54), St. Paul (?) (p. 109) together with *Cardium decoratum* and *C. groenlandicum*.

Pectunculus Lamk.

P. kaschewarowi (nob.) Plate V, fig. 3, a–d.

Testa suborbiculari, radiatim costata; costis 50 applanatis, incrementi striis cellulosis, unbonibus non inflexis minimis, approximatis, dentibus cardinis 28, area ligmenti angusta.

Shape almost circular, only a little extended in width, covered with about fifty flat radial ribs, which are doubly as wide as the intermediate grooves. Numerous cross-lines, thus the intermediate grooves cell-shaped. Hump little elevated, with triangular fields lying tightly together. Each of these hump fields is furnished with eight ribs, which come together in the midline of the triangular field at an angle of 120°. Along the baseline of this triangle, corresponding to the ribs, there are in each bowl sixteen teeth, which become stronger towards the margin. Following these there are six more teeth, which grow in size, so that there are twenty-eight teeth in each bowl, standing on the hinge field, which becomes from inside to outside progressively wider. Shell to 4 m. thick.

Dimensions:

Length	Width	Thickness	
45 m.	47 m.	25 m.	= 1 : 1.15 : 0.55

Baseline of the hump field 19–20 m., height 5 m., hinge field 1.75 m. at the narrowest location.

Occurrence: reportedly on Alaska (p. 28) and St. Paul (p. 109), with the *Venerupis petitii* on Unga (p. 54) and Kodiak (p. 52). *P. sublaevis* from the greensand of Blackdown (Sowerby, Min. Conch. p. Agassiz, p. 492, Tab. 472) has the same outer shape, but a different hinge field. *P. angusticostatus* (Desh. Coq. foss. des environs de Paris, Pl. XXXVI, fig. 20, 21) has more strongly protruding humps and the hump fields lie not so closely together.

The *Nucula*, described below by Girard (Erman, *Archive*, III, fig. 8, a and b), is not available to us, either, unless some very incompletely preserved rock cores from Atka belong here.

Nucula Lamk.

Nucula ermani Girard. Plate V, fig. 4, a, b.

Bowl diagonally oval, but almost as high as it is long, and only shallowly vaulted. The edges uniformly rounded front as well as back. The lunules flat, long drawn out, gradually blending into the posterior rim. The frontal margin more acute. The hinge teeth are large. Each forms an obtuse angle forward. The ones in front are narrower, almost straight. The marginal hollow is beneath, not between, the rows of small teeth at the hump, the posterior of which reaches over the front hollow. Midway between the two larger muscular depressions there is a small muscular depression; but that sits a little higher. The inside of the frontal margin is very weakly, longitudinally striped. The outside is covered with extremely fine lines of growth. They are, however, interrupted several times; and they form flat wrinkles toward the margin. These side stripes are penetrated with narrow lon-gitudinal lines. They begin singly in the hump. Toward the first quarter of their height they part and bring about an inversely V-formed stripe design, which repeats itself severally, side by side, and intertwined, but only in the middle. The frontal and posterior margins are simply grated.

Occurrence: with *C. aleuticum* on Atka (p. 87).

Saxicava Fleuv. de Bellev.

Saxicava ungana (nob.). Plate VI, fig. 1, a–c.

Testa rhomboidali, aequivalvi, umbonibus inflexis distantibus, lateralibus, car-dine edentula, valvis subhiantibus.

Shape even-valved, uneven-sided, rhomboidal, front margin rounded, back rim almost straight, blunted. Hump close to the front end, a little rolled in, and divided from the little bent hinge rim by a wide surface. Hinge rim and lower margin almost parallel.

Dimensions:

Length	Width	Thickness
17 m.	32 m.	13–17 m. = 1 : 1.88 : 0.76–1.

Exterior of ligament long; hinge teeth are missing, front muscular depression deep, the posterior one wider, flatter. From there to the margin the shell becomes abruptly thinner. It gaps in front, and perhaps also a little in back.

The bowl up to 3 m. thick, smoothly or weakly concentrically striped, and seldom well preserved.

Occurrence: Unga (p. 54) at Sakharof [Zachary] Bay and in the vicinity of Illuksky Harbor (?). It most closely resembles *S. arctica*, which is found in living and fossil form (Phillipi, Enum. Molluscorum Siciliae, p. 20, Tab. II, fig. 3). But in our specimen the posterior margin is not so straight. The form of the shell is there-fore less trapezoidal, and the gap less conspicuous. But Phillipi says: "Valvulae mox omino clausae, mox in margine ventrali valde hiantes, mox in latere antico" (cf. also *S. rugosa* from Beauport near Quebec, Trans. of the Geolog. Soc., VI, 1841, P. I, p. 137, Tab. XVII, fig. 7).

Mya L.

Mya crassa (nob.) Plate VI, fig. 2, a–d.

Testa triangulari, antice inflata postice compressa, subangulata.

Plate VI

Fig. 1 a–c *Saxicava ungana*
Fig. 2 a–d *Mya crassa*
Fig. 3 a–c *Mya arenaria*

It is triangular in form, flat in back, inflated in front in the region of the side-ward-placed hump, with mostly clearly indicated margin, running from the hump obliquely forward toward the lower margin. From there to the frontal margin the shell falls off abruptly. Posterior margin more pointed and more gaping than the frontal rim. Hinge aperture: in the left shell a concave tooth of the right shell. Muscle depressions: the posterior one is round, the frontal one oblong with wavy contour. Pallial sinus large (Plate VI, fig. 2, d). Shell thick, often with cross-wrinkles.

Dimensions:

	Length	Width	Thickness	
1)	83 m.	103 m.	53 m.	$= 1 : 1.24 : 0.63$
2)	75 m.	90 m.	50 m.	

> (also 70 m. : 90 : 52 m. Avatcha Lake near Cape Rutmiantsev on the northern part of Port Bodega)

	Length	Width	Thickness
3)	62 m.	80 m.	40 m.
4)	55 m.	75 m.	35 m.

Occurrence: numerous on the SE coast of Alaska [Peninsula] near Pavlof Village (p. 28), and then at Morshovsky [Morzhovoi? or Morozovskii/Cold Bay?] Bay (p. 29), Moller Bay (?) (p. 31), on Kodiak (p. 52), and on Unga (p. 54).

It mostly resembles the living *M. truncata*, which also is found among our petrifacts from Unga (p. 54), Kodiak (p. 52), and Alaska (p. 28 and 29) (Trans. of the Geolog. Soc. VI, London, 1841, P. I., p. 137, Pl. XVII, fig. 5, 6, Canada). It differs from the *M. lata* (Sowerby, Min. Conch., p. 123, Tab. 81) by a less protruding hump, and from *M. intermedia* (l.c., p. 115, fig. 461) by the flat, narrower posterior region.

Mya arenaria L. Plate VI, fig. 3, a–c.

Testa transversim ovata, subaequilaterali, postice subangulata, sinu palliari angustius aperto.

Dimensions:

Length	Width	Thickness	
60 m.	90 m.	35 m	$= 1 : 1.5 : 0.58$
68 m.	110 m.	40 m.	$= 1 : 1.61 : 0.58$

The large tooth in the left shell lies horizontally, its length 20 m., width 15 m. The muscle impressions indistinct (weak), large, and without wavy contour, pallial sinus small. Transverse wrinkles preserved on the rock cores. Occurrence: on Unga (p. 54), Kodiak (p. 52), Alaska (p. 28 and 29), Atka (?) (p. 87).

Mya arenaria var.

Thin-shelled; usually just rock cores. Shape oblong, egg-shaped. Hump almost in the middle and only a little forward, protruding under 120°. The maximum thickness is two-fifths of the width of the shell toward the front, in a ledge which runs from the hump to the margin. Flat toward the back; and the margin extended in that place, and a little dehiscent.

Dimensions at the rock cores:

Length	Width	Thickness	
50 m.	75 m.	23 m.	= 1 : 1.5 : 0.46
24 m.	34 m.	10 m.	= 1 : 1.41 : 0.41
		[Average]	= 1 : 1.45 : 0.43

Muscle depressions simple, small, close to the margins and far from each other. Mantle scar indistinct, pallial sinus distinct generic.

In the left shell a large hollow (concave) tooth, into which are laid two teeth of the right shell. Hump field with indistinct hollows for the outer ligament. Outside covered with fine transverse wrinkles.

Occurrence: Unalaska (p. 70).

Tellina L.

Tellina endentula Brod. et Sew. var. Plate VII, fig. 1, a–c (Middendorff, *Mal. Ross.*, p. 578, and by the same, *Sib. Reise*, Vol. II, part 1, Tab. XXI, fig. 1).

Testa magna, non multum transversa, orbiculari-suntrigona; subaequilaterali, latere antico longiore; subaequivalvi, margine dorsali utrinque valde declivi; area postica inconspicua, dentibus minimis, sublamellosis; sinu palliari congruo, magno, impress. musc. anticae valde approximato, attamen cum hac non confluente, sed denique retroflexo et angulo valde acuto cum impressione palliari sese jungente.

Form egg-shaped, flat, with little, protruding, closely adhering humps; front margin a little drawn out and not so round as the posterior one. Left shell depressed in the region of the pallial sinus.

Dimensions:

Length	Width	Thickness	
55 m.	66 m.	22 m.	= 1 : 1.2 : 0.4
47 m.	55 m.	18 m.	= 1 : 1.17 : 0.38
		[Average]	= 1 : 1.8 : 0.39

Ligament exterior, 26 m. long in larger specimen, and the largest width of the ligament surface 5 m. Hinge teeth in the left shell more developed, but indistinct. Muscular depressions and umbo in the right shell less developed than in the left. The pallial sinus almost parallel to the margin (outline) of the shell, in the left bowl very close to the posterior muscular depression, and downward: long, narrow, and drawn out to a point.

At the lower, peeled-off layers of the shells, a weak longitudinal striping can be observed. The surface is mostly rubbed off; transverse striping, however, is distinct. Thickness of the shell is 2 m.

Occurrences: Alaska, Morshovsky Bay (p. 29), and Moller Bay (p. 31), Unga (p. 54). The variety of this *T. endentula* is especially determined by the fact that the pallial sinus of the right shell never forms a wavy line (as in Middendorff, *Sib. R.*, Tab. XXI, fig. 1). It runs instead analogously on the left shell. It is possible that *T. dilatata*, defined by Girard (Erman, *Archive*, 1843, p. 544, fig. 5a and b), might belong to this species or to the one described in sequel.

Tellina lutea? Gray (cf. Middendorff, *Malacoz.*, p. 578).

Testa magna, calcarea, epidermide vernicosa, decidua tenuissima; oblongo-subtrigona, margine basali subrecto; subaequilaterali, subinaequivalvi, compressiuscula; area postica distincta; dentibus primariis in utraque valva duobus, minutis, altero simplici, altero bicuspidato, alternatim inter se insertis; sinu palliari mediocri, a media impr. musc. post. descendente et denique nonnihil retroflexo, impressionem palliarem mediam fere petentem.

Dimensions:

Length	Width	Thickness		
44 m.	70 m.	15 m.	=	1 : 1.59 : 1.34

The specimens are poorly preserved, but wider and less thick than the preceding *Tellina*. A very small, thin-shelled *Tellina* from Unalaska could also belong to one of the other species.

Occurrences: Alaska (p. 29), Unga (p. 54), Unalaska (p. 70), St. Paul (?) (p. 109), Atka (?) (p. 87).

Astarte Sow.

Astarte corrugata Brown. Plate VII, fig. 2, a–e (cf. Middendorff, *Mal. Ross.*, p. 564, Tab. XVII, fig. 8–10).

Testa vix transversa, cordata, subaequilaterali, crassa; plicis concentricis nullis, vel saltem obsoletissimis; epidermide incrementi striis concentricis striata, in adultis fibrosostriata.

Plate VII

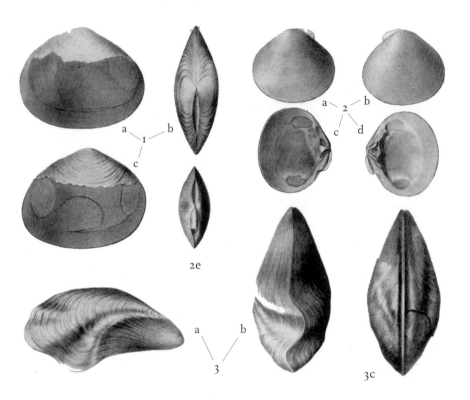

Fig. 1 a–c *Tellina edentula* var.
Fig. 2 a–e *Astarte corrugata*
Fig. 3 a–c *Mytilus middendorffi*

Dimensions:

Length	Width	Thickness		
33 m.	36 m.	16 m.	=	1 : 1.09 : 0.48

These measurements differ but little from those of the living *A. corrugata* of Alaska. Our specimens have turned into chalcedony, and they are often filled with grit.

Occurrence: St. Paul (p. 109).

Mytilus Lamk.

Mytilus middendorffi (nob.) Plate VII, fig. 3, a–c.

Testa subrhomboidali, tumida; valva altera plicis radialibus late et profunde plicato-sulcata; sulco maximo totam fere tumiditatis carinam summam concomitante, paullatim increscente et denique in ultimum marginem ventralem ob plicas hasce valde sinuatum, exeunte; sulcis reliquis duobus multo minoribus, submarginalibus in ultimum marginem dorsalem, rectiusculum excurrentibus; valva altera obsoletissime plicato-sulcata.

Form ham-shaped, pointed, ending with those beaks of the shell that are bent downward or to the left. The beaks of the shell distend but little from each other, perhaps not at all dehiscent. The toothless hinge forms a callosity. The frontal muscular depression is small, near the beaks and the back induration; the posterior one is large and flat.

Dimensions:

Length	Width	Thickness		
45 m.	90 m.	43–45 m.	=	1 : 2 : 1

From the bent (upward bent) back, there runs from the region of the frontal muscular depressions to the lower posterior margin, on one half of the shell, a groove and a tuberosity attached to it (which continues farther downward). On the other half of the shell there is a corresponding tuberosity and groove. Thus, on the lower margin they form two deeply drawn-out fossae. The appearance of groove and swelling is not adherent to a certain half of the shell, in such a way that there would always be observed first the tuberosity, and then the groove. Thus, on the lower margin the first fossa cuts in sometimes to the left, sometimes to the right, and likewise the second one. The shell reaches a thickness of 5 m., and it is covered with undulating wrinkles that run across grooves and tuberosities and grow stronger toward the margin. The black epidermis is occasionally still preserved.

Occurrences: Kodiak (Igatskoi Bay and Cape Tonkii [Ugak Bay and Narrow Cape], (p. 52), and Unga (p. 54).

At Pavlof Harbor (Alaska, p. 28) there is another *Mytilus* (perhaps *Mytilus galloprovincialis*. Philippi, Enum. Moll. Sic., p. 73, Tab. V, fig. 13) which resembles *M. edulis* more closely. But I have not been able to investigate more thoroughly that block which contains this species, and which is kept at the Medical-Surgical Academy at St. Petersburg.

Ostrea L.

Ostrea rudes.

Ostrea longirostris Lamk. var. (Goldf., T. LXXXII, fig. 8, a–c.)
Ostrea testa ovata vel ovato-oblonga, crassa lamellosa, valva superiore convexo-plana, inferiore convexa, umbone elongato recto vel varie inflexoaffixa.

Shape: long, oval, beak bent left.

Dimensions:

Length	Width	Thickness
130 m.	90 m.	35 m.

Shell unfolded, rough, corroded. Lime-stroma very thick, leafy. Lower half of the shell convex, and the hinge groove within it very deep, striped transversely with lateral swellings. Upper half on the shell flat, with an elevation corresponding to the hinge groove 20 m., and the open space between the terminals of the beaks and the closed shell 14 m.

Place of discovery: Unga (p. 54), Alaska (Pavlof Village, p. 28). Furthermore we have "folded Ostreae" (*Ostr. plicata*) from the same place of discovery, and from Unga (p. 54) and Atka (p. 87). Their more exact classification was not possible because of the defectiveness of the specimen. On some of them the fleshy parts have turned completely to calcium carbonate.

Pecten Brug.

Pecten sp. (?)

Dimensions:

Length	Width	Thickness
125 m.	115–125 m.	30 m.

Form almost round. Ears distinct; hinge groove a hollow, weakly transverse-striped conus, with two side grooves in the right shell. The right half of the shell is flatter than the left. Fifteen to sixteen acute ribs with hollow, smooth, wide interstices, which are wider on the left side of the shell than on the right side (the same as *P. solarium* Goldf., XCVI, fig. 7, a, b). The muscular depressions are large and deep, near the rim. Thickness of the shells to 6 m. Weak transverse striping is indicated.

Occurrence: Alaska (Pavlof Village, p. 28)

Crassatella Lam.(?)

From Kodiak (p. 52) or Alaska (p. 28), occurring together with *Mytilus midden-dorffi*, and not more exactly definable.

Venus L.(?)

Impressions in loose tuff of Unalaska (p. 70), and in the clay of Atka (p. 87).

The gastropods from the Aleutian Islands have reached us only in fragments. Among them: *Tritonium (Fusus) anglicanum?* (Transactions of the Geological Society, Vol. VI, P. I, p. 136, Pl. XVI, fig. 1 and 2 from Beauport near Quebec, Newfoundland, Greenland) from Unalaska (p. 70), likewise from Alaska, north of Katmai Bay (p. 26).

Turbo and *Trochus* from Unalaska (p. 70).

Diluvial Remnants

Mammoth bones from Eschscholtz Bay (p. 43). From Elephant Point to Buckland (pp. 44, 45, 46): mandible, tusks, femur, epiphysis, tibia, scapula, os innominatum, os calcis from *Elephas primigenus* and *indicus* (?); head of *Bos urus*, the same as of *B. moschatus*; antlers from *Cervus tarandus* (?); tibia and radius of *Cervus?*; astragalus, metacarpus and metatarsus of *Equus caballus*.

Mastodon ribs, shinbones and tusks of *Elephas primigenus* and *E. indicus* (?) from Cape Nugwulinuk [Black Point], by Voznesenskii (p. 38). Mastodon bones and teeth occur frequently at the coast between Bristol Bay and Norton Sound (p. 35) (Veniaminov, I, p. 105). Furthermore, according to Dr. Stein, they were also found on the Pribilof Islands (p. 109) and finally on Unalaska (p. 69).

Alluvia

Cardium, Venus, Turbo, Murex, Solen, Trochus, Mytilus, Mya, Lepas, Tellina, Asteria in volcanic sand at Cape Espenberg (p. 43), according to Beechey.

Modiola (?) from Shishmaref Bay, by Voznesenskii (p. 41). Horn coral from Amaknak Island (p. 75), and coral (reefs?) at the SW end of Umnak opposite Samalga (p. 76). Without exact indication of the place of discovery, Dr. Stein's essay *(Trudy Mineralogicheskago Obshchestva*, St. Petersburg, 1838, p. 382) speaks of chamites, terebratulites, muricites, helicites, and ichthyolites, which are supposed to be found on the islands.

The remnants of plants are not exactly defined and not well enough preserved to assign them with certainty to the hard coal period, the brown coal period, or to the diluvial or alluvial era. We summarize as follows:

Conifers *(Abies)*, silicified and transformed into brown iron ore, from Kodiak (p. 52), Unga (p. 54), Alaska (p. 28). *Neuropteris acutifolia* (?) from Unga (p. 54).

Dicotyledons *(Alnus)* and conifers *(Taxodium)* from Chugatsk (p. 23), and Unalaska (p. 71) in brown coal and potter's clay. Most of the existing coal strata, as well as the amber-containing clays, belong most likely to brown coal formations (pp. 27, 28, 70, 75).

Explanation of the Plates

Plate IV

Fig. 1, a–d. *Ammonites wosnessenskii. a:* cross section at the level of the orifice; *b:* dorsal view; *c:* lateral view with lobes and saddles obliquely unwound and enlarged (*R:* dorsal line; *DS:* dorsal saddle; *L:* upper lateral lobe; *LS:* upper lateral saddle; *l:* lower lateral lobes; *ls:* lower lateral saddle); *d:* auxiliary lobes and saddles (*NR:* outer umbilicus margin; *NL:* suture line).

Fig. 2, a–d. *Amm. biplex. a:* lateral view; *b:* dorsal view; *c:* cross section of the orifice; *d:* lobes and saddles.

Fig. 3, a–g. *Cardium decoratum. a, b:* right and left halves of the shell of a old individual; *c:* dorsal view; *d:* rock core; *e:* left half of the shell of a young individual; *f:* view of the same from above; *g:* hinge formation.

Fig. 4, a, b. *Cardium groenlandicum. a:* right side of the shell; *b:* view from above.

Plate V

Fig. 1, a, b. *Cardium aleuticum* (Gir.). *a:* left side of the shell from outside; *b:* right side, interior.

Fig. 2, a–e. *Venerupis petitii* (var.). *a:* right half of the shell from outside, with muscular depression, mantle scar, and pallial sinus on the rock core; *b:* left half from outside; *c, d:* hinge apparatus and depressions on the interior of the shell; *e:* view from above.

Fig. 3, a–d. *Pectunculus kaschewarowi*. *a*: right half, exterior; *b*, *c*: right and left halves from inside; *d*: view from above.

Fig. 4, a–b. *Nucula ermani*. *a*: left half, exterior; *b*: right half from inside.

Plate VI

Fig. 1, a–c. *Saxicava ungana*. *a*, *b*: right and left halves of the shell, exterior with muscle depressions on the rock core; *c:* exterior view, dorsal and lateral.

Fig. 2, a–d. *Mya crassa*. *a*, *b*: right and left halves of the shell's exterior; *c*: view from above; *d*: with muscle depressions on the left half.

Fig. 3, a–c. *Mya arenaria*. *a*, *b*: right and left halves of the shell, exterior, *c*: hinge apparatus, etc.

Plate VII

Fig. 1, a–c. *Tellina edentula*. *a*, *b*: left and right halves of the shell, exterior with muscle depressions, mantle scar, and pallial sinus on the left half; *c*: view from above.

Fig. 2, a–e. *Astarte corrugata*. *a*, *b*: right and left halves of the shell from outside; *c*, *d*: right and left halves from the inside; *e:* view from above.

Fig. 3, a–c. *Mytilus middendorffi*. *a*: right half of shell, from outside; *b*: view from beneath; *c:* dorsal view with muscular depressions on the rock core.

APPENDIX II

→→ ←←

Materials for a History of the Voyages and Discoveries on the Western Half of North America and in the Adjacent Oceans

A. Historical Sources of Primary Importance

1. Müller (Gerhard Friedrich): *Sammlung Russischer Geschichte,* Vol. I–IX. St. Petersburg, 1732–1764, 4°. Part III, 1758, contains: "Nachrichten von Seereisen und zur See gemachten Entdeckungen, die von Russland aus längst den Küsten des Eismeeres und auf dem östlichen Weltmeere gengen Japan und Amerika geschehen sind. Zur Erläuterung einer bey der Akademie der Wissenschaften verfertigten Landkarte" [Reports of nautical voyages and ocean discoveries, which were undertaken from Russia along the coasts of the Arctic Ocean and on the Eastern Ocean toward Japan and America. This is a commentary to accompany a map, designed at the Academy of Sciences], 305 pp. [English translation] *Voyages from Asia to America,* etc., second edition, London, 1761 and 1764, 2 vol., 4°. French translation by C.G.F. Dumas, Amsterdam, 1766, 2 vol., 8°. Cf. Coxe.

This is the major source for the earliest Russian voyages of discovery in the Northeast. No mention is made of Spanish or Portuguese voyages. These materials had been collected by Müller in old Siberian archives, especially in Yakutsk (1736). For the second Kamchatka expedition he received reports directly from the same.

For the drafting of his map, Müller used:

1) The map: *Nouvelles découvertes au nord de la mer du Sud, dressée sur les mémoires de Mr. Delisle par Ph. Buache 1750,* and *Explication de la carte des nouvelles découvertes au nord de la mer du Sud par Delisle,* Paris, 1752, 4°.

2). *Nouvelles cartes des d´couvertes de l'Amiral de Fonte, et autres navigateurs par Mr. Delisle,* Paris, 1753, 4°, and *Considerations géographiques et physiques sur les nouvelles découvertes par Mr. Buache,* Paris, 1753, 4°.

Müller's map first appeared in 1754, then in 1758 as *Nouvelle carte des découvertes faites par des vaisseaux russes aux côtes inconnues de l'Amerique septentrionale, avec les pays adjacents, dressée sur les mémoires authentiques de ceux qui ont assisté à ces découvertes et sur d'autres connaissances,* à St. Pétersbourg à l'Académie

Impériale des Sciences. The same is also found in A. L. Schloezer's *Allgemeine nordische Geschichte*, Halle, 1771, 4°, between pp. 390 and 391. In 1773, the map appeared with additions (Pallas, *N.B.* I, p. 275); and in the St. Petersburg Geographic Calendar for 1774, there appeared a short report concerning the newly discovered northern archipelago by Privy Councilor Staehlin: *An Account of the New Northern Archipelago, Lately Discovered by the Russians*, by Mr. J. von Staehlin; translated from the German original. London, 1774, 1 vol., 8°.

2. Fischer (Johann Eberhard): *Sibirische Geschichte von der Entdekkung Sibiriens bis auf die Eroberung dieses Lands durch die Russische Waffen* [Siberian history from the discovery of Siberia until the conquest of that region by means of Russian arms]. St. Petersburg, 1768, 2 parts, 8°. The third and fifth book tell of the discoveries of the NE coast of Asia.

3. Adelung (Johann Christoph): *Geschichte der Schiffahrten und Versuche, welche zur Entdeckung des NO-Weges nach Japan und China von Verschiedenen Nationen unternommen worden, zum Behufe der Erdbeschreibung und Naturgeschichte dieser Gegenden* [History of the nautical voyages and attempts on the part of different nations, which led to the discovery of the NE route to Japan and China for the purpose of geographic descriptions and the natural history of these regions]. Halle, 1768, 1 vol., 4°.

4. Schloezer (J. L. S.): *Neue Nachrichten von denen neuentdekten Insuln in der See zwischen Asien und Amerika; aus mitgetheilten Urkunden und Auszügen verfasset* [Latest reports from the newly discovered islands in the sea between Asia and America, compiled from reported documents and excerpts]. Hamburg and Leipzig, 1776, 8°. Not so much a history of the voyages, as a physical description of the archipelago between Asia and America. Notes on this essay and corrections of a treatise presented to Count Buffon by Pallas in 1777, and published in "Sept Epoques de la Nature," can be found in:

5. Pallas (P. S.): *Neue nordische Beyträge zur physikalischen und geographischen Erd und Völkerbeschreibung, Naturgeschichte und Oekonomie* [New northern contributions to physical geography and ethnography, natural history and economy]. St. Petersburg and Leipzig, 1781, 8°. Vol. I, pp. 273–313: "Erläuterungen über die im östlichen Ocean zwischen Sibirien und Amerika geschehenen Entdeckungen" [Commentaries on the discoveries in the Eastern Ocean between Siberia and America]. A map of the discoveries between Siberia and America to the year 1780 is added. This, too, seems to belong to the "Commentaries."

Pallas's contributions were published as:

Neue nordische Beyträge, Vol. I–IV, 1781–1783, and as *Neueste nordische Beyträge*, Vol. I–III, 1793–1796. The latter also as *Neue nordische Beyträge*, Vol. V–VII.

6. Coxe (William): *Account of the Russian Discoveries between Asia and America, to Which Are Added the Conquest of Siberia*, etc. London, 1780, 1 vol., 4°; 2nd [4th] edit., 1804. French transl., Paris, 1781, 3 vol., 4°; Neuchatel, 1781, 1 vol., 8°. In the main, Coxe made extensive use of Müller; and he adds in the appendix (I, p. 251–266) Krenitsyn's and Levashev's voyage in the years 1768 and 1769.

The following brochure is to be understood as an appendix to Coxe: *A Comparative View of the Russian Discoveries with Those Made by Captains Cook and Clerke; and a Sketch of What Remains to Be ascertained by Future Navigators...by Coxe.* London, 1787, 4°.

7. Busching (Dr. Anton Friedrich): *Magazin für die neue Historie und Geographie*, Vol. XVI. Halle, 1782, 4°, pp. 235–286. "Nachtricht von den Russischen Entdeckungen zwischen Asia und Europa, aufgesetzt von Dr. Pallas und aus dem St. Peterburgischen historisch-geographischen Kalender für das Jahr 1781 übersetzt vom Consistorialrath und Superintendenten Hase" [Reports on the Russian discoveries between Asia and Europe, compiled by Dr. Pallas and translated from the St. Petersburg Historical and Geographic Calendar for the year 1781 by Consistorial Councilor and Superintendant Hase].

8. Forster (Johann Reinhold): *Geschichte der Entdeckungen und Schiffahrten im Norden.* Frankfurt an der Oder, 1784, 1 vol., 8°. *History of the Voyages and Discoveries Made in the North*, translated from the German of John Forster, etc. Dublin, 1786, 8°.

9. Forster (Georg): *Geschichte der Reisen die seit Cook an der NW und NO Küste von Amerika...unternommen worden sind.* From the English original. Berlin, 1791, 3 vol., 4°.

10. Fleurieu (C.P. Claret): Introduction, ou histoire abrégée de la découverte progressive de la côte du Nord-Ouest de l'Amérique, depuis l'année 1537, que Cortès découvrit par mer la Californie, jusqu'en 1791, que le Capitaine Marchand aborda à cette côte par le 53me parallèle. Cf. Marchand (E.), *Voyage autour du monde*. Paris, l'an VI, 4°. Vol. I, pp. i–cxxvii.

11. *Voyage de Humbolt* (A. de) *et Bonpland* (Ai.), *Troisième Partie: Essai politique sur le royaume de la Nouvelle-Espagne.* Paris, 1811, 4°. Vol. I, pp. 328–350: "Un coup-d'oeil rapide sur les côtes du grand Océan, qui, depuis le port de San Francisco, et depuis le cap Mendocino, s'étendent jusqu'aux établissemens russes fondés dans la baie du prince Guillaume."

12. Krusenstern (A. J. von): *Beyträge zur Hydrographie der grössern Ozeane.* Leipzig, 1819, 1 vol., 4°, pp. 85–87, 106, 107, 113–115, 122, 220–235. *Recueil de mémoires hydrographiques, pour servir d'analyse et d'explication à l'atlas de l'océan Pacifique.* St.-Pétersbourg, 1827, 1 vol., 4°, pp. 5–8, 28–112, 401–424. *Supplémens au recueil de mém. hydr. etc.* St. Pétersbourg, 1835, 1 vol., 4°, pp. 98–125, 170.

13. Scoresby jun. (W.): *An Account of the Arctic Regions, with a History and Description of the Northern Whale-Fishery.* Edinburgh, 1820, 2 vol. in gr. 8°. Vol. I, Ch. I and Appendix No. III a, pp. 54–72: "A chronological enumeration of voyages undertaken by the different nations of the world, in search of a northern communication between the Atlantic and Pacific oceans."

14. Berkh (V.): *Khronologicheskaia istoriia otkrytiia Aleutskikh ostrovov . . .* [Chronological history of the discovery of the Aleutian Islands]. St. Petersburg, 1823, v Tipogr. N. Grecha, 1 part, 8°, 169 pp. and Tabl.

Valuable with regard to mercantilist and historical subject matter and as supplement to Busching's *Magazin* XVI. But poor, by contrast, in contributions to the natural sciences. Berkh used mainly the bequest of Shelikhov, in addition to the manuscripts.

15. Baer (K.E. von) and Helmersen (Gr. von): *Beiträge zur Kenntniss des russischen Reiches und der angränzenden Länder Asiens.* St. Petersburg, 1839–1849, 16 vol., 8°.

Vol. IX, 1845: Brief report on scientific projects and journeys toward the knowledge of the Russian Realm [in German].

Vol. XVI, 1849: Peter the Great's contributions to the expansion of geographic knowledge [in German].

16. Stuckenberg (J. Ch.): *Versuch eines Quellenanzeigers alter und neuer Zeit für das Studium der Geographie, Topographie, Ethnographie und Statistik des Russischen Reiches* [Experimental index of sources, early and recent, for the study of the geography, topography, ethnography and statistics of the Russian Empire]. St. Petersburg, 1849, 8°, Vol. I, Section I: Maps, Plans and Monographs, 93 pp., and supplement, pp. 94–142. Cartography of the Russian Realm, pp. 75–89 and pp. 129–132.

The following are worth mentioning as a brief survey of the voyages, but without indication of sources:

Barrow (John): *A Chronological History of Voyages into the Arctic Regions, etc.* London, 1818, 1 vol., 8°. (French: Paris, 1819, 2 vol., 8°.)—"Chronologischer Ueberblick der merkwürdigsten im 18ten und 19ten Jahrhundert in Russland oder von Russland aus unternommenen Reisen" [Chronological survey of the most noteworthy voyages undertaken in Russia or originating from Russia] in *St. Petersburg et Kalender*, 1842 and 1843, in part taken from: *Polnoe sobranie uchenykh puteshestvii po Rossii* [Complete collection of scientific travels around Russia], published by the Imperial Academy of Sciences, St. Petersburg, 1818, 7 parts, 8°.

Falkenstein (Carl): *Geschichte der geographischen Entdeckungsreisen* [History of geographical voyages of discovery]. Dresden, 1828, 5 vol., 8°. Vol. III relevant here.

Lebrun (H.): *Abrégé de tous les voyages au pole Nord, depuis Nicolo Zeno jusqu'au Capitaine Ross (1380–1833)*. Bruxelles, 1837, et Tours, 1841, 1 vol., 12°.

B. Chronological Survey and Index of Sources of the Voyages in the Western Half of North America and in the Neighboring Oceans

1. *Introductory survey of the nautical voyages along the coasts of North America, especially of the attempts at discovering a passage from the Atlantic to the Great Ocean, until the time of Bering*

(cf. Müller, *Sammlung Russischer Geschichte* III; Adelung, *Geschichte der Schiff.;* J. R. Forster, and G. Forster, I; Fleurieu in Marchand's *Voy.*; Humboldt, *Nouv. Esp.*, I; Scoresby, Chron. enum.; O. von Kotzebue, *Reise um die Welt von 1815–1818*, Vol. I, pp. 1–73).

Iceland and Greenland became known during the ninth (A.D. 861) and tenth (A.D. 982) centuries. In the fourteenth century N. and A. Zeno most likely advanced even farther. But not until the fifteenth century was New Foundland visited, first supposedly by the Portuguese (1463 John Vaz Costa Cortereal), but with certainty by the Englishmen John and Sebastian Cabot[136] (1494–1497). They advanced to a latitude of 67°30′ N in their attempt to find a NW passage to India. After Central America had become known, the Portuguese Gaspar and Miguel Cortereal continued with these attempts (1500–1505). And the former thought that in Hudson's Bay he had found a navigable passage from the Atlantic to the Pacific Ocean. He named it with the Christian name of one of these brothers, the Strait of Anian. With Magellan's first voyage around the globe (1519–1522), the Spaniards[137] had advanced westward to Molucca, without overcoming the earlier difficulties in the way of communicating with East India, which the Venetians and the Portuguese had also faced (Vasco da Gama, 1497). Therefore, the Spaniards looked about for a closer and less dangerous way.

Emperor Charles the Fifth sent Estevan Gomer de Corunna in 1524 to find a passage from the North of America to the Moluccan Islands. But this endeavor failed, and Gomer returned to Toledo in 1525 (J. R. Forster, p. 513; S. Miguel Venegas's *History of California*, p. 124). Englishmen and Frenchmen had just as little success with similar attempts from 1524 until 1536. Therefore, from then on the attack was undertaken from the west coast of North America. Cortez, the conqueror of Mexico, had heard of the attempts made by Cortereal. It is said that in 1537, after California had been discovered by his companions in 1534,[138] and after he himself had visited it in 1536, Cortez sent Francisco Ulloa with three ships to find the Anian Passage from the west coast of America. Exact reports about the success of that voyage are lacking. Following Cortez, the viceroy, Antonio de Mendoza, outfitted two expeditions, one overland under Francisco Vasquez Coronado, the other under Francisco de Alarcon. Both had the task of finding the Anian Strait, and to explore the coast up to 53° N. Lat. As far as anyone knows, Alarcon got only as far as 36° N. Lat. (cf. Antonio Herrera, *Description de las Indias*, Ambères, 1728, fol.; also in Latin language, Amsterdam, 1622. Also, Jo. de Laet, *Novus orbis, seu Americae utriusque descriptio*, Antwerp. et Lugd. Bat. ap. Elzevir, 1633, fol.).

The French (Jacques Cartier, 1540) continued to look along the east coast of America for that famous passage, and the Portuguese (F. M. Pinto, 1542, in Adelung's *Geschichte der Schiffahrten*, p. 447) advanced along the east coast of Asia as far as Japan. At that time, 1542, the Portuguese Juan Rodriguez Cabrillo was sent from the west coast of America. He reached 37° N. Lat. (Punta de Anno Nuevo) and died on the third of January, 1543, on St. Bernardo Island. His helmsman, Bartolomeo Ferrelo, continued the voyage and discovered Cape Medocino between 40° and 41° N. Lat., from which point onward to the Port de la Navidad (Capo das Nawados, Cape Blanc, Cape Oxford, according to Vancouver at 43° N. Lat.) everything was a continuous landmass (*Nouv. carte d. découv.*, and Humboldt, *Nouv. Esp.*, I, p. 329, after manuscripts of Don Antonio Bonilla and M. Casasola in Archivo General de Indias, at Madrid).

1555. Martin Chake's voyage belongs to the realm of fables.

1556 or 1557. The monk Andreas Urdanietta did, according to Salvatierra's testimony, find a thoroughfare.

1576–1578. Frobischer's three experimental voyages, attempting to find a passage from the NE coast of America, were unsuccessful. The English also made attempts along the west coast:

1577. Sir Francis Drake's voyage. He sailed around Cape Horn; and in 1579 he took possession of the land above California in a harbor under 38°30' N. Lat. Reports on his voyage we find in: *Expedito Francisi Drake in Indias Orientales, anno 1557*. Leiden, 1588, 1 vol., 4°. *The World Encompassed*. London, 1628, 4°. The author of this brochure is Admiral Drake's ship's preacher by the name of Francis Fletcher. A second edition from 1652. Cf. G. Forster, I, p. 18; Fleurieu in Marchand's *Voy.*, p. vii; and *Nouv. carte des découv.*; also James Burney, *Chronological History of the Discoveries in the South Sea...*, London, 1803–1817, 10 [5] vol., I, p. 343.

1582. Francisco Gali (Franz Gualle) was ordered by the King of Spain to investigate whether it is true that a passage exists east and north of Japan, by which the South Sea is connected to the sea in the north of Asia. Gali reached 57°30' N. Lat., and deduced with certainty from the currents that between the continents of New Spain and Tartary or Asia a canal or strait is to be found (*de Canto. Decad.* 10, lib. 5, cap. 3; *Routier de Linschoten* cap. 54, and Humboldt's *Nouv. Esp.*, I, p. 330, etc.).

1583. The Englishman, Adrian Gilbert, and 1585–1587 John Davis undertook three voyages along the NE coast of North America. The latter reached 73° N. Lat., and believed he really had found the passage.

1586–1588. Thomas Candish, also Cavendish, first voyage with three ships from Plymouth, to New Spain and the Philippines.

1588. It is claimed that Lorenzo Ferrer Maldonado sailed through the Strait of Anian from the Atlantic into the Pacific Ocean. Cf.: *Memorias sobre las observaciones astronomicas por Don Josef Espinosa y Tello,* Madrid, 1809, 2 volumes in 4°. French by Wallenstein (cf. Malespina, 1789) section 4.—Mémoire de Don Ciriaco Cevallos. *Recherches faites dans les archives de Seville, par Don Augustin Cean (Juan). Introduction historique au voyage Galiano et Valdez,* pp. 49–56 and 76–83 (Humboldt's *Nouv. Esp.,* I, p. 329)—*Mémoire d'un voyage du Maldonado* par Amoretti, Plaisance, 1812—*Bibliographie universelle* par C. A. Waldenaer, Paris, 1826–1831.—O. von Kotzebue's *Riese um die Welt, 1815–1818,* Vol. I, pp. 26, 39, 40.—Barrow, *Chronological History of Voyages,* London, 1818, Appendix II, pp. 24–48: "A relation of the discovery of the Strait of Anian, made by Capt. Lorenzo Ferrer Maldonado in the year 1588."—Scoresby's Chron. enum., Edinburgh, 1820.—*Nouvelles Annales des voyages* par Eyriès et Malte-Brun, Paris, 1821, Vol. XI, pp. 1–28: "Voyage de Maldonado par Lapie avec Carte." Lapie's attempt to prove the existence of Maldonado's passage, and similar attempts by other authors, seemed to be futile in the light of recently obtained information about that part of America which lies opposite Asia. But it is surprising that no one thought of moving Maldonado's voyage farther northward, according to his own testimony; thus he would have sailed past Kotzebue Sound and through the Bering Strait. Lapie has Maldonado emerge from Norton Sound.

1591–1593. Candish (Thomas), second expedition (cf. 1586). Candish remained unsuccessful and died on the return voyage (*Bibliographie universelle,* Paris, 1813, Vol. VII, p. 11).

1592. Juan de Fuca, a Greek whose real name was Apostolos Valerianos, set out from Acapulco Harbor and advanced reportedly far above 47° and 48° N. Lat. The inlet between the southern part of Vancouver Island and the mainland has retained his name. Nevertheless, the report from that voyage seems to consist more of fiction than of fact. Cf.: Luke Fox's *North-west Fox.* London, 1635, 4°, p. 163–166.—*The Principal Navigations, Voyages, Traffiques, and Discoveries,* etc., by Richard Hakluyt, London, 1598, 1 vol., folio, and *Hakluytus Posthumus or Purchas, his Pilgrimes,* London, 1625, Book IV of the third Part, pp. 849–852, Michael Lock's apocryphal stories of Juan de Fuca and of his supposed strait.—*Traité des Tartares par Bergeron,* 4°, Ch. 21, p. 125.—*Nouv. carte de découv.*—G. Forster and Fleurieu in Marchand's *Voy.*—Mémoire de Don Ciriaco Cevallos (cf. 1588 [above]).

1596. Sebastian Viscaino's, or Vizciano's, first voyage from Acapulco; Island Mazatlan in New Galicia, and Port San Sebastian (42° N. Lat.). He investigated the coast for a distance of more than one hundred miles northward from California (Humboldt, *Nouv. Esp.*, I, p. 330).

1598. Marquis de la Roche's voyage along the west coast of North America, with the intent of establishing colonies (Scoresby's Chron. enum.).

1598. Olivier van Noort's voyage in the Great Ocean (Fleurieu in Marchand's *Voy.*, Vol. I, p. xii).

1602 and 1603. Vizciano's second voyage (see 1596). Under 36°44' N. Lat., the beautiful harbor Monterey is discovered the second time, and in 41°30' N. Lat. Cape Mendocino, and in 43° Capo Blanco (cf. 1542). The reports concerning the Strait of Martin de Aguilar are exaggerated and wrong. (Torquemada, *Monarchia Indiana*, Madrid, 1723; Humboldt's *Nouv. Esp.*, I, p. 330. The learned cosmographer Enrico Martinez published thirty-two maps in Madrid, based on the voyages of Viscaino.)

1602–1616. The Englishmen George Weymuth, James Hall, Henry Hudson, Thomas Button, Gibbons, Robert Bylot, and William Baffin make vain, but better known attempts at finding a passage from the NE coast of America.

1619. Danish attempts (Jens Munk).

1631. Attempts of the Englishmen Luke Fox, Thomas James. Then there followed a pause of some duration (until 1688). The only voyage known during this time was:

1640. The voyage of the Spanish and Portuguese Admiral Bartholomeo de Fuentes, or Fonte. Cf.: *Nouv. carte des découv. de l'Amiral de Fonte, et autres navigateurs…, avec leur explication par Mr. Delisle*, Paris, 1753, 4°, and *Considérations géograph. et physiques*, etc., par Mr. Buache, Paris, 1753, 4°.— *Observations critiques sur les nouvelles découvertes de l'Amiral Fuentes*, etc., par Robert de Vaugondy, fils, Paris, 1753. *Journal Historique, Mémoire pour l'histoire des sciences et des beaux arts, Journal des Savans, Journal économique pour l'année 1753.* Müller's *R. G.* [Russian history] III, p. 71—G. Forster, I, pp. 22–35.—Fleurieu in Marchand's *Voy.*, Vol. I, pp. xv–xxx.—Mémoire de Don Ciriaco Cevallos (cf. 1588) — *Nouvelles Annales de voyages*, par Eyriès et Malte-Brun, Vol. XI, 1821, pp. 28–56: Voy. de Fonte, par Lapie, with map, whereon is outlined the route taken by Fonte and one of his captains, "Pedro Bernarda." Lapie attempts to harmonize the voyages of Bernarda and Maldonado, and he indicates that the former, too, made the northern passage from the Atlantic to the Pacific Ocean (cf. 1588).

1668. A voyage was undertaken by Zacharie Gillman from Quebec into the Polar Sea.

All the later attempts (e.g., 1741 and 1746, two voyages by Christoph Middleton, William Moor, Francis Smith, and Henry Ellis) to get around the polar coast from the east coast of North America are hereafter recorded only when they are connected with expeditions along the western coast.

1b. Introductory Survey of the Voyages Along the North and Northeast Coast of Asia to Bering

The attempts of Englishmen after 1553 (Willoughby, Chancellor, Nennet, Poole, and others) and the Dutch after 1594 (Barents, and others) to advance along the northern coast of Asia into the Pacific Ocean, led by way of Novaia Zemlia to the Ob' estuary and resulted in the discovery of Spitzbergen. The major benefit for the British and Dutch companies on Spitzbergen was the whaling, the butchering of sea lions, and fishing. Attempts to advance farther were not undertaken. Thus the expanded knowledge of the coast of Northern Asia happened thanks to the Russians, who began in 1636 to navigate from Yakutsk down the Lena River and into the Arctic Sea (Müller's *Sammlung Russ. Geschichte*, III, p. 6). In 1639 they advanced as far as the Sea of Okhotsk (Ritter's *Asien*, II, p. 601).

From the south the Dutch (1600) and the English (1613) reached Japan along the east coast of Asia. In 1643 the East India Company sent off captains Martin Heritzoon van Vriez and Heinrich Cornelius Schäp, from the harbor on Ternäte Island. They sailed northward along the Kuril Islands and reached 47°08′ (Adelung, *Geschichte der Schiffahrten*, pp. 473–496, and J. R. Forster, p. 487). The only other reported voyage from that time which might be of interest, is that of Jean de Gama. Only through the 1649 map of the Portuguese cosmographer, Texeira, did the details of the story of that traveler become known. According to it there are 10°–12° north of Japan, under 44°–45° N. Lat., a multitude of islands, and a coast that runs out toward the east (cf. on Buache's map: *Terre vue par Jean de Gama, Indian en allant de la Chine à la nouvelle Espangne*, and *Considérations géogr. et phys.* par Buache, p. 128.—Müller, *R. G.*, III, pp. 195 and 288.—Adelung, *G. d. Schiff.*, pp. 496–498).

At about the same time, Russian hunters under Isai Ignat'ev, 1646, and 1647–1649, Fedor Alekseev from Kholmogory, Cossack Semen Ivanov Dezhnev,[139] and Gerasim Ankudinov, expanded their excursions from the Kolyma River to the mouth of the Anadyr River; also Stadukhin and other promyshlenniks, all of whom heard the rumor of a large island in the east (cf. Müller's *R. G.*, II, p. 6; Adelung's *G. d. Schiff.*, pp. 507–517; J. R. Forster; and G. Forster). But not until the expeditions of Atlasov, from 1696 to 1700, are some reports received about Kamchatka and the Kuril Islands, after the last futile voyages of Zacharie Gillman (1668 from Quebec) and John Wood (1676 by way of Novaia Zemlia), whose main assignment it was to find a passage to India. But the Kuril Islands were not reached until 1711—in the course of the Cossack regime and the disorders on Kamchatka—by the mutineers Danila Antsyferov and Ivan Kozyrevskii (Müller, *R. G.*, III, p. 73; [Adelung] *G. d. Schiff.*, pp. 519–525, after Strahlenberg's NE portion of Europe and Asia, p. 431; Storch's *Portrait of Russia*, Vol. V, p. 166—Storch's *Russland unter Alexander I*, Vol. I—Weber's *Verändertes Russland*. III, p. 159; and Baer and Helmersen, *Beitr.*, Vol. XVI, p. 34). In 1713 the governor of Kamchatka at that time, Kolesov,[140] ordered the occupation of two of the Kuril Islands. But larger voyages on the Sea of Okhotsk were not undertaken, until Swedes started them under the command of the Cossack Sokolov, as a result of an order by Peter the Great, in 1716 (Müller's *R. G.*, p. 102; Adelung, *Geschichte der Schiffahrten.*, p. 542; Baer and Helmersen, *Beitr.* XVI, p. 34). In 1719, Peter the Great sent the surveyors Ivan Evreinov and Fedor Luzhin[141] with secret orders to the Kuril Islands. The former delivered to the tsar a map of the same (l.c., and Müller, *R. G.*, III, p. 109). Death prevented Peter the Great from implementing the expedition of Bering, which he had ordered. But the Empress Catherine I executed his plan. The success of Bering's voyage brought one of the most important geographic discoveries of all times. And since in the course of these, the Aleutian Islands and the NW coast of America had been visited for the first time, we begin with them the actual chronological survey and bibliography of the voyages to the regions covered in our discourse.

2. Voyages from Bering to Cook, 1725–1776

1725–1730. First Kamchatka Expedition led by Vitus Bering, with Chirikov and
    Spangberg [Spanberg].

Berkh, V., *Pervoe morskoe puteshestvie rossiian, predpriniatoe dlia resheniia geograficheskoi zadachi: Soediniaetsia li Aziia s Amerikoiu? i sovershennoe v 1727–1729 godakh, pod nachal'stvom Flota Kapitana 1-go ranga Vitus Beringa* [First Russian sea voyage undertaken to resolve the geographical problem: Is Asia joined to America? and carried out in the years 1727–1729, under the command of Fleet Captain First Rank Vitus Bering]. St. Petersburg, 1823, 8°. Also cf. *Zhizneopisaniia pervykh rossiiskikh admiralov* [Biographies of the first Russian admirals], Pt. II, pp. 219–226 (Bering's Life). Müller, *R.G.*, III, pp. 112–138. Adelung's *G. d. Schiff.*, pp. 550–559. Baer and Helmersen, *Beitr.* XVI, pp. 39–96.

On the fifth of February, 1725, Bering left St. Petersburg and in 1728 he plied the coast of Kamchatka. He reached the land of the Chukchi, observed the easternmost point of the same, and discovered St. Lawrence and the St. Diomede Islands. He noticed that under 67°18' (a point not known to date, due north of

Cape Herzfels, Cape Serdtse Kamen') the coast turns westward in an acute angle. Thus, farther northward the land completely disappeared from his sight. On the sixteenth of August, as the result of a counsel of war, he turned about from the just-mentioned point. He returned by land, and reached St. Petersburg again on the first of March 1730.

1727–1731. The expedition of Cossack Colonel Afanasii Shestakov and Dmitrii Pavlutskii consisted of 400 Cossacks, helmsman Jacob Hens, second helmsman Ivan Fedorov, geodesist Mikhail Gvozdev, the miner Herdebol, and ten sailors (cf. Adelung, *G. d. Schiff.*, pp. 550–568). The major result of this expedition was that Gvozdev reached the mainland of America in 1730, between 65° and 66° N. Lat. He also visited the islands which were temporarily named after him: Gvosdef Islands. However, they had already been discovered and named the St. Diomede Islands by Bering. In the same year (1730), Melnikov, who had been sent to the land of the Chukchi in 1725, had received reports at Chukotskoi Nos about the large island or land across from that cape. Thus, they had been known prior to the Gvozdevian discovery.

1733–1743. Second Kamchatka Expedition. The most important members of this expedition, other than its commander, commodore Veit (Vitus) Bering, were: Chirikov, Spangberg [Spanberg], Walton, Gmelin, Müller, Steller, Fischer, Delisle de la Croyère (Louis), Krasheninnikov, and Krasil'nikov. This very extensive expedition resulted in the following publications:

1) J.G. Gmelin, *Reise durch Sibirien von dem Jahr 1733 bis 1743*. Göttingen, 1751–1752, 4 parts, 8°. French edition, Paris, 1767.

2) De l'Isle de la Croyère (Joseph Nicolas): *Explication de la carte des nouvelles découvertes au nord de la mer du Sud, par Ph. Buache 1750*. Paris, 1752, 4°.

3) *Lettre d'un officier de la marine Russienne à un seigneur de la cour de Berlin, concernant la carte des nouvelles découvertes au nord de la mer du Sud, et le mémoire qui y sert d'explication publié par M. De l'Isle à Paris en 1752*. Trad. de l'Original Russe, 1753. According to von Baer (in *Beitr.* XVI, pp. 58–59), authored by F. G. Müller. The same letter in the *Nouvelle Bibliotheque Germanique* 1752, 4°, Vol. XIII, p. 52; and finally, corrections of this publication in Müller's *R. G.*, III, and Fischer's *Sib. G.*, 1768.

4) *Stepana Krasheninnikova opisanie zemli Kamchatki*. St. Petersburg, 1755. Cf. also: *Polnoe sobranie uchenykh puteshestvii po Rossii*…[Full collection of scholarly travels about Russia], pt. I, St. P., 1818. Stepan Krasheninnikov's description of the land Kamchatka, translated from the English excerpt of Mr. Jeffry, by Johann Tobias Kohler. Lemgo, 1766, 4°. *Histoire et description du Kamtchatka par Kracheninnikow, prof. de l'Ac. des Sc. de St. Pétersbourg*. Traduit du Russe. Amsterdam, 1770,

2 vol., 8°. The German translation is just as useless as the English one. Best is the French translation in Abbé Chappé d'Auteroche: *Relation d'un voyage en Siberie, fait par ordre du Roi en 1761*. Edition de Cassini. 2 vols. in 4°, avec Atlas in folio. Paris, 1767.

5) G.W. Steller's *Beschreibung von dem Lande Kamtschatka*, etc., published by J. B. S. (Scherer). Frankfurt and Leipzig, 1774, 1 vol., 8°. Cf. also Pallas, *N. B.* V, pp. 129–236. Sequel VI, pp. 1–26. "Steller's Reise von Kamtschatka nach Amerika," in *Allgemeinen Historie der Reisen ʒu Wasser und ʒu Lande*, vol. XX. Leipzig, 1771, pp. 357–361: About the islands located between Kamchatka and America.

The assignments and successes of this expedition are treated in general in Müller's *R. G.*, 1758, Vol. III, pp. 138–305; in Adelung's *Geschichte*, 1768, pp. 568–705.; in Phipps's *Reise nach dem Nordpol, unternommem im Jahre 1773* [Voyage to the North Pole, undertaken in the year 1773], from the English into German, Bern, 1777, 1 vol., 4°, with appendix, pp. 1–304, by Samuel Engel: *Neuer Versuch über die Lage der nördlichen Gegenden von Asien und Amerika...*[New attempt to explain the position of the northern regions of Asia and America, and probes of a passage through the North Sea to India, etc.] (French edition, Paris, 1775, 2 vol., 4°). Presently at press there is a new edition of this voyage in von Baer's and Helmersen's *Beitr.*, Vol. XVI, by the academician A. von Baer, who had available to him several, previously unused, manuscripts. Of exclusive interest to us here is only Bering's and Chirikov's assignment, to search for America with two ships, which were not to part from each other. The voyages of these seamen were described and charted on a map by the astronomer Krasil'nikov. The map is filed at the Admirality in St. Petersburg in manuscript form. It is supposed to be more exact than that of Müller from 1758, etc. Bering's and Chirikov's voyages are furthermore critically appraised in the logs of Cook (third voyage), La Pérouse, Vancouver, Marchand (Fleurieu), and Belcher, but especially in Krusenstern's *Beyträge ʒur Hydrographie...*, 1819, and *Mem. Hydr.*, 1827.

Bering and Chirikov set sail from Avacha Harbor (presently Peter-Paul's Harbor [Petropavlovsk]) on the fifth of June, 1741, the former with Steller on the *St. Peter*, and the latter with Delisle de la Croyère on the *St. Paul*. They held a SSE course until the twelfth of June (46° N. Lat.), then a northerly course to 50° N. Lat., and thereafter an eastern course until the twentieth of June, when in storm and fog they lost eye-contact under 49° N. Lat. After five weeks of sailing, Bering reached the coast of America between Cape St. Hermogenes and Cape St. Elias (Cape Suckling in the region of the Copper River) on the twentieth of July. He cast anchor at an island opposite that of cape (Kayak or Wingham). And after staying only long enough to take on fresh water, he turned about. During his return voyage he sighted St. Dolmat (Four Summit) Mountain and Kodiak. On the second of August, 1741, he sighted in 55°32′ (corrected between 55°46′ and 56°), Tummanoi or Fog Island, which Vancouver later named after Chirikov. But the true name used by the natives, Ukamok, was made known by Sarychev. On the fourth of August he touched upon the Eudokeyevian [Evdokeev or Semidi] Islands (55°45′, corr. 56° to 56°05′ N. Lat.), and from the twenty-ninth of August until the fourth of September he remained in the area of the Shumagin Islands, which he named Shumagin after the sailor who was buried on Nagai Island. From then on, Bering was unable to leave his quarters because of illness. Lieutenant Waxel took over the command. On the twenty-fourth of September he sighted St. John's Mountain in 54°27′ N. Lat. (corr. 54°45′; Khaginak or Shishaldin). ([Note from p. 424]: Adelung indicates 51°27′ N. Lat. and 21° of longitude from Avacha. But that is obviously a typographical error. It should read 54°27′. The longitude is totally wrong.) On the twenty-fifth of October he anchored near St. Markiana (Amchitka, 50°50′ N. Lat., corr. 51°15′ N. Lat.). On the twenty-eighth of October at noon he sighted St. Stepan (Kiska, 51°55′ N. Lat., corr. 52°10′–12′) and three small islands east of there, perhaps Chugul and Khvostov, and on the twenty-ninth of October at 10 a.m. St. Abraham (the Semichi Islands, 5°31′ corr. 52°43′–50′ N. Lat., which he took for one island). From then on, it seems, a constant course of WNW and NW ¼ W was held. With a western course one ought to have reached Peter-Paul's Harbor in a few days, had the latitudes not recorded an error of 15′ to 25′ throughout, sometimes even ½° to ¾°. Therefore, they believed they were too far south of Avacha Bay. On November 6, 1741, Steller and probably the majority of the crew believed they were approcaching Kamchatka. They approached instead the NE or N side of Bering Island in 55°05′ N. Lat. (corr. 55°20′–22′ N. Lat.). There Bering died on the eighth of December. Those forty-four who remained of the original seventy-six, did not leave under sail until the sixteenth of August 1742, reaching Avacha on the twenty-sixth of August.

Concerning the voyage of Aleksei Chirikov, we refer to an edition of the same, now being prepared for publication by Lieutenant-Commander A. P. Sokolov. It is assumed to date, that Chirikov touched upon the American coast on the fifteenth of July at about 57° N. Lat. He turned around on the twenty-seventh. On the first of August he also sighted Chugatsk, on the fourth of September the Fox Islands, on the ninth S. Amchitka, and on the twentieth some land in 51°12′ N. Lat. But whether it was St. Theodor and St. Julian, Attu and Amchitka, we cannot tell. On the seventh of October he entered Peter-Paul's Harbor. We add here only a few remarks, which come mainly from our famous seaman Krusenstern (who probably had access to all the manuscripts at the Admirality). Krusenstern states (*Beytr. zur Hydr.*, p. 228), that Chirikov had sighted Cook's Mt. Edgecumbe earlier (15–21 July, 1741), calling it Mt. St. Lazarus (G. Sv. Lazariia). Even without access to Chirikov's ship's logs and Krasil'nikov's map, this statement seems very acceptable. It is because, according to Müller's map from 1758, Pallas *N. B.* I, p. 269, La Pérouse 4° II, p. 223, and others, Chirikov sent a rowboat ashore under 57° N. Lat., under Trubitsyn's command. This has not been confused with the boat which was sent out farther north with fourteen men under the leadership of Abraham Dement'ev and Sidor Savelev, who abandoned it. On a map published in the Depot of the General Staff in St. Petersburg, 1801, Mt. Edgecumbe is even called Cape Chirikof; also Trubitsyn [Trubitsin Point] and Quadra's Cape San Bartolome (58°12′–13′ N. Lat., according to Malaspina 55°17′ N. Lat., 133°36′ W. Long.) on an island at the entrance into Bucarelli Harbor is called by that name [Point Chirikoff]. "It is not easy to discern," states Krusenstern (l.c.), "why a name given by the Spaniards in the year 1775—a name respected by La Pérouse and Vancouver—is to be supplanted by another one, which already exists not far from there, unless it has happened as the result of excessive patriotism. It might have been believed that Cape Chirikof, wherever it might be located, would constitute the southern border of the Russian possessions on the coast of America. Less patriotic, but historically closer to the truth, is probably this: to maintain that Chirikov sighted the coast of America not as far south as 55°17′ N. Lat." In addition to Vancouver, who named Ukamok Island after Chirikov, La Pérouse has given him a lasting memorial with Chirikof Bay (Christian's Sound), and with the southern point of Baranof Island. The name Cape Chirikof, which is found on Sarychev's chart, is older than Colnett's Cape Ommaney, older, too, than Malaspina's name, Punta Oeste de la Entrada del Principe, which had to be abandoned as impractical. If Chirikov was indeed at Kruzof Island, then it is not very patriotic that the names given by the first discoverers, St. Lazarus or Trubitsyn, were forgotten or not maintained, as we notice at the Hydrogr. Dept. at the Naval Ministry in St. Petersburg on the 1847 Mercator map of the region between the islands Baranof and Kodiak. There we see again Colnett's indication of Cape Ommaney and Christian's Sound, and only at the southern coast of Kruzof Island, opposite Mt. Edgecumbe, the island St. Lazarus. This inconsideration is partly due to Capt. Lisianskii, who manifested a lack of knowledge of earlier voyages during the christening of Kruzof Island and several bays in the region, or else a measure of conceit. Chirikov, Quadra, Cook, Vancouver, Portlock, and Dixon had sighted St. Lazarus, and they called the latter the island to which it belongs, Pitt's Island. Then the agents of the [Russian] American Company introduced the island's true name: Sitka. Lisianskii's Shelikof Bay is Vancouver's Port Mary, etc. In this regard we want to mention that the settlement of the American company was at first on the present Kruzof Island, called Sitka at that time. And after it had been destroyed by the Kolosh in 1804, it was moved to the present location as New Archangel (Novo-Arkhangel'sk). With the same, the name Sitka seems also to have been transferred to Baranof Island. For the island which has been entered on Russian maps as Kruzof Island since 1813, the following names have applied in chronological order: St. Lazaro, San Jacinto, Edgecumbe, Pitt, Sitkha, Kruse. A dictionary of synonyms for the names of the islands, bays, promontories and mountains of those regions would be a commendable achievement. But from among the many designations, the local names presently in use, and the names of the first discoverers should be established once and for all, and the rest assigned to oblivion.

1743–1744. Emel'ian Basov's first voyage. He wintered on Bering Island.[142]

1745–1746. Second voyage by the same, and Nikifor Trapeznikov. Bering Island, Copper Island, and two islands located farther south.

1745–1747. Mikhailo Nevodchikov and Iakov Chuprov. Three new islands, among them Attu and Karag. Reports about Shemijia (Amlia). In Falkenstein, III, p. 135, the name Novosiltsov is probably confused with Nevodchikov (cf. also Pallas, *N. B.* III, p. 279).

1746–1748. Andreian Tolstykh (1) and Fedor Kholodilov. Bering Island.

1746–1749. Andrei Vsevidov. Exact reports not available. He probably stayed the winter on Copper Island.

1747 Basov's third voyage.

1747–1749. Ivan Rybinskii and Stepan Tyrin remained on the Near Islands for two years.

1747–1749. Afanasii Bakhov and Novikov. According to Coxe and Berkh, pp. 15 and 16, Seamen [Semeon] Novikof and Ivan Bakhof. According to Busching's *Mag.* XVI, p. 249, the time was from 1748 to 1749. The reports in Busching are older, and Berkh does not provide sources for his assertions. Berkh says that after the shipwreck on Bering Island they went northeasterly. On that journey they sighted land. And had they continued in their determination, they would have discovered the mainland of America. But since, because of fog, they lost the mentioned land from sight, they turned around. Busching states: "Finally they built a new small vessel, 17.5 *arshin* long, with which they first undertook a voyage northeastward, where they had imagined an unknown land, which, however, they failed to find. Therefore they sailed for Copper Island." Stuckenberg, in his *Hydrographie d. Russ. Reiches* [Hydrography of the Russian Empire], Vol. II, p. 708, mentions the statement of Berkh. But he fails to subject it to the critique it deserves.

1747–54. Zhilkin's and Studentsov's voyage. Bering Island.

1749–?. E. Basov's fourth voyage. Basov died in 1754.

1749–1750. Nikifor Trapeznikov and Cossack Sila Shivyrin. An unknown island.

1749–1752. Andreian Tolstykh (2). Bering Island and three additional islands, one of them Nevidiskov.

1749–1752. Rybinskii and Tyrin's second voyage. Near Islands.

1750–1752. The voyage of the Cossack Vorob'ev; calamitous.

1751–1754. Emel'ian Iugov's voyage. Bering Island and Copper Island. On the latter he died.

1752–1757. Aleksei Druzhinin. Shipwreck on Bering Island.

Now there follow three simultaneously outfitted expeditions, to visit new islands and the Bol'shoe Zemlitse, by which name was meant new islands and the mainland of America at that time (as Berkh states, p. 25). Only Serebrennikov's voyage was genuinely successful.

1753–1754. Andrei Mikhail[ov] Serebrennikov's ship under the command of Maksim Lazarev and the Archangelite Bashmakov. Several new islands, one of which is supposed to lie opposite Khatyrskii Nos. Three others are supposed to surround this one. The Fox Islands (Umnak) were, as Berkh indicates, probably not visited by him, because the Russians exported only arctic foxes until the year 1762 (cf. Veniaminov, I, p. 115). Erman states (in III, p. 35) that the eastern Fox Islands were discovered as early as 1750.

1753–1757. Fedor Kholodilov's vessel visits Bering Island and an unknown Aleutian island.

1754–1757. Seamen [Semeon] Krasil'nikov's ship visited Bering Island, one unknown island, and Copper Island.

1754–1757. Cossack Durnev. Islands: Ataki, Ayataku, and Shemiyae. Reports from the islands Ibiyae, Riksa, and Olas, which were thought to be located toward the east.

1755–?. Petr Iakovlev, mining inspector, is sent to Copper Island by the government. Cf. Pallas, *N. B.* II, pp. 302–308: "*Kurze Beschreibung der Kupferinsel, ein Auszug aus J's Bericht*" [Brief description of Copper Island, an excerpt from Y's (Iakovlev's) report].

1756–1759. Andreian Tolstykh's third voyage. He returns with the first extensive reports about some Aleutian Islands: Atak, two smaller, adjacent islands, and Iviyae Island (see Durnev).

1758–1762. The Muscovite merchant Ivan Nikiforov outfits a ship under the command of Stepan Glotov, resident of Iarensk, and the Cossack Savin Ponomarev. They remained on Unimak from September 1, 1759, until May 23, 1762. Ponomarev delivered to the government a map of the Aleutian Islands, charted by himself and Totemian merchant [merchant from Tot'ma] Petr Shishkin. On it are drawn eight islands east of Unalaska.

1758–1761. Ivan Zhilkin's ship, led by the Cossack Ignatii Studentsov. Calamitous. Bering Island and Near Islands.

1758–1761. Dmitrii Paikov, Cossack Sila Shevyrin [Sava Shavyrin], Aleksei Druzhinin, and Semen Polevoi sail on a ship outfitted by Nikifor Trapenznikov. Bering Island, Atka, and Amlia, or Atakh (Goreloi), and Amlakh, then Ssitkino (Sitkin).

1759–1761. Rybinskii's ship, led by Andrei Serebrennikov and the sergeant Basov. Near Islands and Krugloi (Agattu).

1759–1762. The merchants Postnikov from Shuisk, Krasil'nikov from Tula, and Kul'Kov from Vologda outfitted a ship under Cherepanov from Tot'ma, which seems to have only visited the Near Islands.

1760–1762. Bechevin's ship under Sergeant Pushkarev. Notorious voyage (cf. Cook and Coxe). Atakh [Atka?], Amlia, Siguam, Unimak, Unga, Unalaska or Alaska. According to Veniaminov (I, p.116), Bechevin's ship, under the command of Ponomarev, set out from Atkha, reached Alaska in 1761, and wintered in Protassof Bay on the northern side of the Strait of Isannakh [Isanotski Strait]. Veniaminov does not believe, as Berkh does, that part of the crew stayed on Unga at the same time, but rather, that as a result of the universal uprising of the inhabitants of Alaska, the crew fled to Unimak and remained not far from the ship.

1760–1764. The fourth voyage of Andreian Tolstykh from Selenginsk, with Petr Vasiutinskii and Maksim Lazarev. This is the most important among the voyages undertaken by the promyshlenniks until that time. It brings news of the island Agayae or Kayag, and the six Andreanof Islands: Kanaga, Chetchina, Tagalak, Atkhu, Amlag (Shemiyae), and Atakh (cf. Shelikhov's voyage of 1783–88). The latter group of islands was named after Andreian Tolstykh. But Berkh says in his, *Khronol. istoriia Aleut. ostrovov*, p. 55, that it had already been discovered earlier by Bashmakov and M. Lazarev in the years 1757 and 1758. If there was no mistake in printing here, and the years 1753 and 1754 are [not?] actually meant, then the just-named promyshlenniks must have made another voyage in 1757 and 1758.

1760–1763. Chebaevskii's ship (without further reports).

1761. Abbé Chappé d'Auteroche: *Voyage en Siberie, fait par ordre du Roi en 1761* (Paris, 1767, 2 vol., 4°, avec Atlas in fol.); *Antidote ou examen, du mauvais livre superbement imprimé, intitulé: Voyage en Sibérie, fait par ordre du Roi en 1761* (Amsterdam, 1771, 1 vol., 8°); *Voyage en Californie pour l'observation du passage de Vénus sur le disque du soleil, le 3 Juin 1769, par M. Chappé d'Auteroche* (Paris, 1772, 1 vol., 4°).

1761–1766. Expedition of four ships, three of which did not return. The reports are very sketchy; cf. Veniaminov, I, pp. 118–131.

　　1) 1762. Aleksei Druzhinin with thirty-four men, only three of whom return (six, according to Veniaminov; among them Bragin, who stayed

the winter of 1765 on Kodiak. Cf. Sarychev, II, p. 37). Atakh, Umnak, Unalaska Captain's Harbor at Ubienna River, which flows from the south.

2) Ivan Korovin. Bering Island, Unalaska, Umnak. After they lost their ship, they returned in 1765 with Solov'ev (cf. 1764–1765).

3) Helmsman Medvedev's ship is lost near Umanak, according to Veniaminov, p. 118.

4) The ship outfitted by the merchants from Lal'sk, Terentii Chebaevskii, Vasilii and Ivan Popov, and the Solikamsk merchant Ivan Lapin, commanded by Stepan Glotov, returns in 1766. Copper Island, Umnak, Kodiak, Aktunak or Akutanaka, Saktunu (?), Unalaska. Glotov stayed the winter of 1763 on Kodiak (Sarychev, II, p. 37).

1764–1765. Ivan Maksim[ovich] Solov'ev's voyage. Unalaska.

1764–1766. The Totemian merchants [merchants from Tot'ma] Grigorii and Petr Panov outfit two ships. Merchant news only.

1764–1768. Ship's Lieutenant Sind is sent out by Catherine II, to investigate the islands in the Polar Sea between America and Asia. He discovered St. Matthew Island (cf. Coxe, *Acc. Russian Discoveries*, p. 300).

1765. Two ships are readied, the *Peter* and the *Paul*, by the Solikamsk merchant Lapin, the Velikoustiugian merchant Shilov, and the Tula armorer Afanasii Orekhov. The ship *Peter* stood under the command of Andreian Tolstykh. He, however, perished on this, his fifth voyage, at Cape Shipunskii with his entire crew (except for three men). The ship *Paul* was commanded by Afanasii Ocheredin. It returned in 1770. Atkhu or Achak, Umnak, Uliaeg, Akutan, and Akun (Pallas, *N. B.* I, p. 276; Sarychev, II, p. 37).

1766. Seamen [Semeon] Krasil'nikov of Tula readied a ship, commanded by Sapozhnikov. It visited the Fox Islands, so called because on them there are black, dark brown, blue and red foxes, while on the Rat and Near Islands only *pesty* (ice foxes) were encountered.

About the eight voyages made by promyshlenniks during the next four years, we have only merchant reports (cf. *Khronol. istoriia Aleut. ostrovov*, pp. 82–86).

1767–70. The ship *St. Peter and Paul* is readied the second time by Grigorii and Petr Panov. Later on, outfitted the third time, it fails to return.

1767–70. The ship of Ivan Popov. It makes three voyages, the last one until 1772.

1767. The vessel *St. John the Baptist*, sent out by the Greek, Peloponisov, and by Popov.

1768–73. The ship *Nikolai*, from the company: Ivan Zasypkin, armorer Afanasii Orekhov from Tula, and Ivan Mukhin from Tobol'sk.

1769–73. The vessel *St. Andreas* [*Adrian*], outfitted by Peloponisov and Popov, is wrecked at Fort Udskoi.

1769–73. The vessel *St. Prokopius* [*Prokopii*], belonging to Matvei Okoshinikov from Volodga, and Prokopius Protodiakonov.

1770–74. The vessel *Aleksandr Nevskii*, belonging to Vasilii Serebrennikov from Moscow.

1770–75. The *St. Paul*, sent out by Orekhov, as well as Lapin and Shilov.

1767. The Velikoustiugian merchant Vasilii Shilov, whom the Empress Catherine had ordered in 1764 to come from Okhotsk, handed to the Collegium of the Admiralty a map of the Aleutian Islands; cf. 1765, and *Khronol. istoriia Aleut. ostrovov*, p. 70.

1768–69. The voyage of Captain Krenitsyn and Levashev, in Coxe: *Account of the Russian Discoveries*, etc., Appendix I, pp. 251–266, with map. Translated verbatim in Pallas, *N.B.* I, pp. 249–272, "Account of the sea voyage accomplished in the years 1768 and 1769 by order of the Russian Empress, under the command of Captain Krinitsin and Lieutenant Levashef, from Kamchatka to the newly discovered islands and to Alaska or the mainland of America." Cf., furthermore, Busching's *Mag.* XVI, p. 269, and Sarychev, II, p. 20.

1769–70. Two packet boats, *San Carlos* and *San Antonio*, commanded by Don Vicente Vila and Don Juan Perez, and a troop commanded on land under Gaspar de Portola are sent by the Viceroy of New Spain to claim the NW coast of America up to Cape Mendocino. Fort Monterey is constructed. Cf. Fleurieu in Marchand's *Voy.*, pp. lv–lx; La Pérouse, Vol. I, pp. 123–125; Diario historico de los Viages de Mar y Tierra hechos al Norte de la California. Should the Juan Perez, mentioned above, be someone other than he who made the voyage in 1774 on the corvette *Santiago* (*San Yago*), then Nootka Sound was already discovered before 1774.

1769. The Danish Baron von Uehlefeld claims to have reached the Pacific Ocean by way of Hudson's Bay. Scoresby's Chronolog. enum.

1769–1772. Samuel Hearne dispatched to the northern polar sea under contract to the Hudson's Bay Company. The Coppermine River is discovered. G. Forster, I, p. 99, and introduction, by the same, to Cook's *Third Voyage*, p. 33; Lebrun, pp. 106–124; Fleurieu in Marchand's *Voy.*, Vol. IV, Pl. I. This voyage proved indisputably that the mainland of America extends at least as far north as 68° N. Lat., and that consequently no passage farther south can be thought of.

1770–1775. Ivan Solov'ev's voyage. Pallas, *N.B.* III, pp. 326–334, and *St. Petersburg Zeitung*, 1782: "Excerpt from the log of a sea voyage which Ivan Solov'ev undertook in the years 1770–75, to the Alaska Peninsula, which belongs to the mainland of America." The Krenitzin Islands become more accurately known as a result of this voyage.

1772–1776. The voyage of the apprentice helmsman D. Bragin. This is the same Bragin who, with Korelin and Kokovin, survived the Druzhinin expedition of thirty-four men in 1761–64. Pallas, *N. B.* II, pp. 308–324: "Report of the Peredovshik, Dmitri Bragin, of a sea voyage of one year's duration, begun in 1772, to the islands located between Kamchatka and America." P. 320: Tanakh (Takovania?) with volcano and hot springs, Semisopuchnoi (Unyaek!) with volcano, p. 322: Island Sitignak, probably between Amchitka and Ayugadakh or Krissy-Ostrov, with one fire-belching mountain and hot springs.

1772–1778. Helmsman Potap Zaikov's voyage with the *St. Vladimir*. Copper Island, Attu, Agattu (Kruglyi?), Umnak, Unimak, Sülatis, Sannakh, Ungin, Kodiak.... Cf. Pallas, *N. B.* III, pp. 274–288: "Excerpt from the travelogue of the Russian Helmsman Zaikov about a ship's voyage to the mainland of America, with map." In Berkh's *Khronol. istoriia Aleut. ostrovov*, the voyage is mentioned according to the excerpts by Pallas. The voyages of Krenitsyn, Levashev, and Bragin are not mentioned.

1772–1780. The ship belonging to the Tot'ma merchant, Aleksei Kholodilov, is commanded by Dmitrii Polutov. Unalaska and Kodiak (Igatsky Bay), Berkh, pp. 94–99.

1772. A ship is dispatched by the Tot'ma merchants Petr and Grigorii Panov. No further reports.

1773. The vessel belonging to the merchant Fedor Burenin from Vologda, by the name of *St. Yewell* (*Yevel*) [*Evpl*], leaves the harbor of Nizhne-Kamchatsk. Without further reports, just like the next voyages which follow.

1774. The baidara belonging to the Kamchatkan merchant Ivan Novikov visits the Near Islands for the second time (1772–1774).

1774–1778. The ship, *St. Prokopius*, sails for the second time. It had been outfitted by the merchants Protodiakonov and Okonishnikov. Merchant reports only.

1774. The vessel *St. Paul*, belonging to the merchant Osokin, founders.

1774. Juan Perez and his helmsman Esteban Jose Martinez leave San Blas on the corvette *Santiago* (formerly called *Nueva-Galizia*) on the twenty-fourth of January. On the twentieth of July they discover Charlotte Island (La Marguerita), anchor for the first (?) time (cf. 1769–72) in Nootka Sound (San Lorenzo); and they winter from the twenty-seventh of August, 1774, in Monterey. After the manuscript of Fray Juan Crespi and Fray Tomas de la Penna, in Humboldt's *Nouv. Esp.*, I, p. 331; Roquefeuil, II, p. 174.

1775. Investigation of the NW coast of America by the Spaniards Don Heceta, Don Juan de Ayala, and de la Bodega y Quadra, under orders of the viceroy Don

Francisco Bucarelli e Ursova [Antonio Bucareli y Ursua]. Pallas, *N. B.* III, pp. 198–273: "Log of a voyage, undertaken in the year 1775 for the investigation of the coasts continuing northward beyond California by the Second Pilot of the Spanish fleet Don Francisco Antonio Maurelle on the Royal Galiot, *Sonora*, commanded by Don Juan Francisco de la Bodega, from the English language of Magellaen, or after M. Daines Barrington's *Miscellanies*, London, J. Nichols, 1781, 4°." Cf. Fleurieu in Marchand's *Voy.*, I, pp. xliv–lvii; La Pérouse's *Voyage autour du monde*, par Milet-Mureau, Vol. I, pp. 125–144; Espinosa: *Memorias sobre las observaciones astronomicas*, etc., Vol. II, part 2, or Wallenstein's French translation; and finally Pallas, *N. B.* I, pp. 269–271. They discovered the estuary of the Columbia River, the Punto de la Trinidad, the island Quadra (Vancouver) with Nootka Sound, this for the second time, and the beautiful Bucarelli Harbor on Prince of Wales Island, which is surrounded by seven volcanos, the summits of which are snow-covered and throw out fire and ashes (Humboldt, *Nouv. Esp.*, I, p. 332). This report might be based on an error, because in Pallas, *N. B.* III, p. 247, only the following is mentioned: "We found (24 August) the air temperature very comfortable, which presumably seemed to have its reason in the presence of some mighty fire-mountains (volcanos), the fire of which we could observe at night, in spite of the large distance." And in Pallas, *N. B.* I, p. 271: "The harbor Bucarelli opens an excellent view inland, where there are mighty volcanos visible." Cf. also J. R. Forster, pp. 521–524; La Pérouse, p. 136. If Chirikov was not the one who, already in 1741, sighted Mt. San Jacinto, naming it St. Lazarus, then these men were the first to touch Edgecumbe or Kruzof Island (cf. Fleurieu in Marchand's *Voy.* I, p. l).

3. Voyages from Cook to the Most Recent Times

During the time of Bering's expeditions, the English, too, continued their unflagging attempts to proceed to the Pacific Ocean from the NE side of America. For the solving of this task, Parliament appropriated the sum of 20,000 pounds sterling in 1743, when the results of the second Kamchatka Expedition could be expected. But when all attempts failed, attention was once again directed to the NW coast of America. Cook's voyage constitutes for this reason, and because of its importance, a new worthwhile chapter.

1776–1779. James Cook's third voyage: *A Voyage to the Pacific Ocean,…Performed Under the Direction of Captains Cook, Clerke, and Gore,…in the Years 1776–1780*, vol. I and II written by Capt. J. Cook, vol. III by Capt. James

King, accompanied by a folio volume of maps, charts, portraits, views, etc. (London, 1782, 1784 and 1785, 4° and 8°, 3 editions); *The Original Astronomical Observations Made in the Course of a Voyage to the Northern Pacific Ocean, in the Years 1776–1780*, by Capt. James Cook, Lieutenant King, and Mr. William Baily (London, 1782, 1 vol., 4°); *Troisième voyage*, transl. from English by M.D. ..., 8 vol. (Paris, 1785, 8°); *Des Capt. Jacob Cook's dritte Entdeckungsreise in das stille Meer und nach dem Nordpol hinauf waehrend der Jahre 1776–1780, nach den Tagebuechern der Capt. Cook, Clerke, Gore, King u. Anderson*, from the English language [into German] by Georg Forster, 2 vol. in 4°, (Berlin, 1787): Vol. II, Cape Foulweather, Nootka Sound, Mt. Edgecumbe, Mt. Fairweather, St. Elias, Cape Suckling, Kaye's Island, Iliamna, Halibut Island, Unalaska, Unella [Unalga], Akutan, Unimak, Bering Strait, Cape Lisburn, Burney Island (78°49′). Charles Clerke continues the investigation of the Arctic Sea north of Bering Strait after Cook's death, but he does not advance farther than 70°33′. Captains Pikersill and Jung, who were sent to meet Cook by way of Bering's Bay, did not get across the Arctic Circle.

1776–1779. The ship, *St. Aleskandr Nevskii*, belonging to the Tot'ma merchants Grigorii and Petr Panov, visits the Aleutian Islands, but detailed reports are lacking (Berkh, p. 99).

1776. We hear for the first time of the Ruelian [Ryl'sk] merchant Grigorii Shelikhov, who came from Kiakhta to Okhotsk. There he outfitted the vessel, *St. Paul*, together with the Kamchatkan merchant Luka Alin. The ship returned in 1780.

1777–1781. Voyages of a second ship, which was readied by these and other merchants; and then yet another ship, 1777, 1778, and 1779 (cf. Berkh, pp. 99–106). The other six expeditions up to 1781, which Berkh mentions, are of interest to merchants only. Among the members of the same were also Jerassim [Gerasim] Gregoriev Izmailov and Ivan [Iakov] Sapozhnikov, whom Cook met at the beginning of October, 1778, on Unalaska (cf. Cook's *Third Voyage*, [transl.] by G. Forster, vol. II, p. 165).

1779. On February 11, de la Bodega y Quadra undertook a second voyage of discovery along the NW coast of America, accompanied by his compatriot, Ignacio Arteaga, with the corvettes *La Princessa* and *La Favorita*. They advanced a little beyond Cape St. Elias, and turned back on the twenty-first of November, 1779 (G. Forster, I, pp. 46 and 47; and La Pérouse, I, pp. 324–342: Extrait de la relation d'un voyage fait en 1779, par don François-Antoine Maurelle, etc.). La Pérouse obtained Maurelle's manuscript on Arteaga's voyage in Manila. In a letter to Fleurieu (Avacha, the tenth of September, 1787), he writes: "Je ne vous envoie pas leur carte générale, parce qu'en vérité elle nuirait plutôt au progrès de la géographie qu'elle ne pourrait y être utile. Ont-ils voulu nous tromper, ou plutôt ne se sont-ils pas trompés eux-mêmes? Quoi qu'il en soit, ils n'ont vu la terre qu'auprès du port Bucarelli et à l'entrée du port du Prince-Guillaume." Cf. Fleurieu in Marchand's *Voy.*, vol. I, pp. lxiv–lxv; and Humboldt, *Nouv. Esp.*, I, p. 333. Only now can we determine how unjust was La Pérouse's assessment of Maurelle, and how regrettable it is that the maps remained unrecognized. The Port San Yago is the harbor north of Cook's Cape Hinchinbrook. Arteaga anchored thereafter at Montague Island. From there he proceeded to the Barren Islands (Bezplodnye ili Peregrebnye, unfruitful islands), so named by Cook. One of these he called Isla de la Regia (59°08′ N. Lat., that is an error of a mere 10′–15′). From that point Arteaga saw in clear weather a volcano, NW 7° W and at 20 miles' (*lieues*) distance. He estimated its elevation higher than the peak of Teneriffe. It was completely covered with snow (first of August, 1779). Nearby in WNW 8°W and 15 miles' distance, another high mountain was observed, on which no snow was noticed. And finally WSW, in 13 miles' distance there were two others, on which there was still snow. The former two mountains were most likely Iliamna Volcano and Augustine Mountain, or maybe the High Mountain (Uyakushatch) [Redoubt Volcano] and Iliamna, because the location of the Isla de la Regia cannot be exactly determined. One of the two others is the Four-Summit-Mountain [Fourpeaked] (cf. also Espinosa, *Memorias sobre las observaciones astronomicas*, etc., in the French edition by Wallenstein, P. II).

1781–1789. The Yakutsk merchant Lebedev-Lastochkin readied the *St. George* under the command of helmsman Gerasim Pribylov, who discovered the islands St. Paul and George, naming them the Zubov Islands. This name, however, was later changed to Pribilof Islands (cf. Count Aug. Benyakovsky's [Beniowsky's] journey through Siberia to Kamchatka and to Madagascar, German edition by G. Forster, Leipzig, 1792; Sauer in Billings, p. 246; *Syn otechestva* 1821, No. XXVII; Berkh, *Khronol. istoriia Aleut. ostrovov*, 107; Veniaminov, I, pp. 131, 132).

1781. Three other endeavors are undertaken from Siberia:

   1781–86. The *St. Paul* was dispatched for the second time by Shelikhov and Alin.

   1781–86. The *Alexius* [*Sv. Aleksei*] outfitted by the Tot'ma merchant Aleksei Popov.

   1781. The *St. Aleksandr Nevskii* readied by Orekhov, Lapin, and Shilov. It is not known when the ship returned. In Berkh, pp. 108, 109, we find only merchant news concerning these three expeditions.

1782. The Siberians sent only one ship, the one which belonged to the Irkutsk merchant Iakov Protasov from Nizhne-Kamchatsk. It returned in 1786. The only reports are about hunting booty.

1783–1785. Three ships under the Pilot Potan Zaikov (cf. 1772–78), who knew Cook's map and sailed into Prince William Sound (Berkh, *Khronol. istoriia ostrovov*, pp. 111–114).

1783–88. Shelikhov's voyage. *Rossiiskago kuptsa…Grigor'ia Shelekhova pervoe stranstvovanie s 1783 po 1787 god…*[First voyage of the Russian merchant Grigorii Shelekhov from 1783 to 1787…]. St. Petersburg, 1790, 8°, 172 p., and *Ross. kuptsa Gr. Shelekhova prodolzhenie stranstvovaniia v 1788 g.* [Continuation of the voyage of the Russian merchant Grigorii Shelekhov in 1788], St. Petersburg, 1792, 95 p. Or [in German] *Grigori Schelechofs…erste und zweyte Reise von Ochotsk in Sibirien, durch den oestlichen Ocean nach den Kuesten von Amerika, in den Jahren 1783–89*, translated from the Russian language by J. Z. Logan. St. Petersburg, 1793, l vol., 8°. The same, in part, in Pallas, *N. B.* VI, pp. 165–249, and in Busse's *Journal für Russland*, 1794, Vol. I.

1783–89. Assistant Helmsman Stepan Zaikov has command of the ship *St. Paul*, which belongs to the Yakutsk merchant Lastochkin. He loses the ship near the islands St. Paul and St. George (Berkh, pp. 109–111).

From now on we see an abatement of the eagerness of Siberian merchants for trade endeavors, because the return is no longer so abundant; and the numerous calamities are discouraging. Stepan Zaikov, for instance, lost one-third of his entire crew on his last voyage. In 1785 a company was founded under the name: "King George's Sound Company." It was intended to erect a settlement in Nukta [Nootka], resembling the settlement of the New Holland Company.

1785–87. James Hanna's two voyages of trade and discovery. The first originated in Typa (on the Canton River). He reached Nootka Sound in August 1785 and discovered Fitzhugh Sound at 51°15′ N. Lat. His second voyage from Macao to Nootka Sound is unimportant. In May of 1787 he returned to Macao (Fleurieu in Marchand's *Voy.*, p. xc, and G. Forster, p. 53).

1785–88. George Dixon's and Nathaniel Portlock's voyage for the English Trade Company of Nootka Sound [King George's Sound Company] to the NW coast of America and around the world. Cf. Fleurieu in Marchand's *Voy.*, Vol. I, pp. xcv–c. *A Voyage Round the World*, etc., by Nathaniel Portlock (London, 1789, 2 vol., 4°). *A Voyage Round the World*, etc., by George Dixon (London, 1789, 2 vol., 4°); French edition, Paris, 1789, 2 vol., 8° or 1 vol., 4°. G. Forster, I, p. 55, and Vol. II: *Der Capitaine Portlock's und*

Dixon's Reisen nach der NW-Küste von Amerika, published by Dixon, translated with commentaries by G. von Forster; Vol. III, pp. 1–165: Diary of the ship's captain, Nathaniel Portlock.

1785–94. Joseph Billings's and Fedor [Gavriil] Sarychev's voyage. *An Account of a Geographical and Astronomical Expedition...performed...by Commodore Joseph Billings in the Years 1785–1794*, narrated by Martin Sauer, secretary to the expedition. London, 1802, 3 vol., 4°. French edition, Paris, 1802, 2 vol., 8°. German edition, *Geograph. astronom. Reise nach den nördlichen Gegenden Russlands auf Befehl der Kaiserin Catharina II, in den Jahren 1785–94, unternommen von Capt. J. Billings und nach den Originalpapieren herausgegeben von Martin Sauer, Secretair der Expedition*, from the English language, with copper plates and maps, Berlin, 1802, published by Oehmigke, 8°.

During all the long years of Billings's expedition the latitude of not one single location was astronomically fixed. But then, nothing more could be expected from the former assistant of the astronomer Bayly (who had accompanied Cook). Thanks to Sauer, the voyage became one of the most significant with regard to natural history.[143]

Puteshestvie Flota Kapitana Sarycheva po severovostochnoi chasti Sibiri, Ledovitomu moriu i Vostochnomu okeanu...pri...ekspeditsii...Billingsa [Voyage of Fleet Captain Sarychev over the northeastern part of Siberia, the Arctic Sea, and the Pacific Ocean...with the Billings Expediton]. St. P., 1802, 2 parts, 4° v tipografii Shnora. Included: Atlas of fifty pages in folio, and the "Merkatorskaia karta severovostochnoi chasti Sibiri, Ledovitago moria, Vostochnago okeana i severozapadnikh beregov Ameriki. Sochinial flota Kapitan Sarychev" [Mercator map of the northeastern part of Siberia, the Arctic Sea, the Pacific Ocean, and the northwestern shores of America, by Fleet Captain Sarychev]. On this map are charted the voyages of the following ships:

1) Voyage of the vessel *Iasashna*, under Sarychev, from the upper Kolyma river into the Arctic Sea, 1787.
2) Voyage of the *Slava Rossii* (Glory of Russia) under Billings, from Okhotsk to Peter-Paul's Harbor, 1789.
3) Voyage of the same vessel, 1790.
4) The same, 1791.
5) Return voyage of the same, under Sarychev, to Unalaska.
6) 1792, voyage of the *Chernyi Orel* (Black Eagle), under Sarychev, to Peter-Paul's Harbor, and from there, under Captain Hall, to Okhotsk.

Gawrila Sarytschew's...achtjährighe Reise in NO-Sibirien, auf dem Eismeere und dem nördostlichen Oʒean [Gavrila Sarychev's eight-year voyage in N.E.

Siberia, on the Arctic Sea, and in the Northeastern Ocean]. Translated from the Russian language (into German) by Johann Heinrich Busse, with copper plates. Two parts, 8°, Leipzig, 1805–1806. Sarychev's work compensates by thoroughness and diligent surveying particularly of the Aleutian Islands, for the shortcomings of that of Billings. He furthermore published: *Puteshestvie Kapitana Billingsa...* Vitse-Admiralom Gavriilom Sarychevym [Voyage of Captain Billings...by Vice Admiral Gavriil Sarychev]. St. P., 1811, 1 vol., 4°, s kart [with maps].

1786–87. On the twenty-second of December, the Empress Catherine the Great issued orders for the organization of an expedition into the Eastern Great Ocean, in order to assert her right to the regions discovered by Russian argonauts ("Dlia okhraneniia prava nashego na zemli, rossiiskimi moreplavateliami otkrytyia"). As a result of this Ukaz, the Admiralty College readied five vessels: *Kholmogory* with forty-two pieces of heavy cannon, *Solovki* with twenty, *Sokol* and *Tarukhtan* with sixteen each, and the transport ship, *Smelyi*, with fourteen. Capt. of the First Rank Grigorii Ivanovich Mulovskii was appointed commander of the expedition and of the ship *Kholmogory*. The *Solovki* was commanded by Capt. Second Rank A. Kireevskii; the *Tarukhtan* by Commander Joachim von Sivers; the *Sokol* by Prince Dmitrii Trubetskoi, and the *Smelyi* by Carl von Grewens. In order for the journal to be written in acceptable style (*chistym stilem*), Mr. Stepanov, educated at Muscovite and English universities, was appointed Secretary and natural scientist upon the recommendation of J.R. Forster of Cook Voyage fame. He received 5,400 rubles for his mobilization and the journey from Vilna to St. Petersburg, an annual salary of 3,000 rubles, and a lifetime pension of 1,500 rubles. In case of his death during the voyage, this amount was to go to his widow. In the opposite case the widow, or the children until their full age, were to receive 750 rubles. Forster also suggested the appointment of Dr. Med. and Chir., Sommering, and of the astronomer Bayly (cf. Cook's voyage). And he promised to provide four design artists as well. Professor Pallas, who was consulted, was on this occasion appointed maritime historian with a salary of 750 rubles, by an Ukaz from December 31, 1786. Thereafter five officers and a number of helmsmen were thoroughly trained in the use of the instruments and astronomical observations by Professor Inokhodtsov at the observatory of the Academy of Sciences. In addition to the most thorough preparations concerning food, medicine, clothing, and the like, different items suited for trade with the wild peoples and with the Japanese were purchased. Then, "for the purpose of confirming the right to the regions so far discovered, and yet to be discovered by the Russians," there were cast 100 cast-iron eagles without inscription or date, and 100 depicting the portrait of the empress with Latin and Russian inscriptions and the dates, 1789, '90, and '91. Finally, there were minted 100 gold coins with eyelets, and 10 without eyelets, 400 silver coins with, and 30 without, 600 copper coins with, and 60 without eyelets, as well as 500 iron coins. A more detailed instruction ordered the commander to proceed to the NW coast of America with two ships, and to take into possession the regions from Nootka Sound to Chirikov's landing point. With two other ships he was to chart exactly all the islands from Japan to Kamchatka; and those from Matmai to Cape Lopatka were to be formally annexed to the Russian empire as possessions. The armed transport ship was to sail straight to Okhotsk. But when, in the autumn of the year 1787, everything was ready, and the expected return of the Empress from the Crimea took place, the longed-for declaration of war on the part of the Porte [Turkey] also took place. And now the Empress suspended her plan, by an order dated October 28, in order to make the ships available for service in the Mediterranean Sea. More detailed information about these preparations can be obtained from *Zapiski, ucheniia komandira Morskago Ministerstva* [Notes, studies of the commander of the Naval Ministry], pt. XV, 1840, pp. 361–367; and the *Zapiski Gidrograficheskago Departamenta Morskago Ministerstva* [Notes of the Hydrographic Dept. of the Naval Ministry], pt. VI, 1848, pp. 142–192. From them we discern that the great Empress had taken care to be active for Russia from that side as well. She did not neglect the scientific objectives alongside her political plans. It is not impossible to assume that she had intended to correct in this way the erroneous selection of Billings for the expedition of 1785. The thought that the expedition merely served as a pretense, seems not probable.

1786–88. Captain William Peters was sent by the East India Company on business, leaving Bengal on the twentieth of March, 1786. On the twenty-eighth of July he arrived at Peter-Paul's Harbor. During the continuation of his voyage to the NW coast of America, he suffered shipwreck on Copper Island. Cf. Shelikhov, p. 61. Fleurieu in Marchand's *Voy.*, Vol. I, p. xc; *Journal historique du voyage de M. de Lesseps...du Kamch. en France*, Vol. I, note, p. 10.

1786. Englishmen Lawrie and Gise, of the East India Company, sail from Bombay to King George's Sound (Nootka Sound). They survey the coast to 60° N. Lat. ([Prince] William Sound), and once again discover Charlotte Island (cf. Fleurieu, in Marchand's *Voy.*, Vol. I, p. xciv; and G. Forster, I, p. 54).

1786. Capt. Meares's and Tipping's voyage from Bengal. On the twenty-second of July, 1787, they were in Prince William's Sound, Port Etches (cf. 1788, Capt. Meares's second voyage).

1786. Barclay's voyage. He left Ostende in November, and in August, 1787, arrived in Nootka Sound. Fleurieu in Marchand's *Voy.*, Vol. I, pp. c and ci; Dixon's *Voyage*, pp. xx and 232; Meares's *Voyages*, p. iv; G. Forster, I, p. 56.

1786–88. Capt. James Colnett and Charles Duncan are dispatched by the Nootka Sound Company (cf. Meares's *Voyages*, p. lv; G. Forster, I, p. 57; Fleurieu in Marchand's *Voy.*, I, pp. ci–cii).

1786–88. Jean Francois [Galaup] de la Pérouse's voyage. He surveys the coast from St. Elias to Monterey and discovers Charlotte Island. Lesseps and Pérouse part company in Kamchatka in 1787. Lesseps returns by land with diaries and maps (cf. *Journal historique du voyage de M. de Lesseps*, Paris, 1790, 2 vol., 8°; *Voyage de la Pérouse autour du monde...* par Milet-Mureau, 4 vol., 4° Paris A° IV (1797), 8°, English translation: London, 1799, 4 vol., 4°; *Voyage de la Pérouse, rédigé d'après ses manuscripts originaux, suivi d'un appendix par M. de Lesseps*, Paris, 1831, 1 vol., 8°). La Pérouse perished with his ship.

1787. According to Shemelin, II, p. 332 (cf. Krusenstern's voyage, 1803–6), a certain Capt. Berill touches on the Russian colonies.

1787–90. The ship *Zosima and Savvatii* is sent out to the Aleutian Islands by the Irkutsk merchant, Iakov Protasov (Berkh, p. 115).

1787–93. The galiot [*Sv.*] *Georgii* makes a similar voyage (l.c.).

1788. Don Esteban Martinez and Don Gonzalo Lopez de Haro set sail on the eighth of March from Port San Blas with the frigate *La Princessa*, and with the packet boat, *San Carlos*. They visit Chugatsk Bay, Cook's Inlet, Kodiak (Kikhtak), Shumagin (Unga), Unimak, Unalaska. On the fifth of December they returned (cf. *Diary of Capt. Douglas*, p. 290; G. Forster, II, p. 44; Fleurieu, in Marchand's *Voy.*, I, pp. cxvi–cxxiii; Humboldt, *Nouv. Esp.*, I, p. 334: "Reconocimiento de los quatro establacimientos Russos al Norte de la California, N-Mexico hecho en 1788").

1788 and 1789. Capt. Meares's second voyage with William Douglas, James Colnett, and Robert Hudson. The construction of a settlement in Nootka Sound is intended. But this is prevented by the Spaniards (Martinez, 1789). *Voyages Made in the Years 1788 and 1789, from China to the North West Coast of America, to which are prefixed, an introductory narrative of a voyage performed in 1786, from Bengal, in the ship Nootka; observations on the probable existence of a North West Passage; and some account of the trade between the North West Coast of America and China; and the latter country and Great Britain*, by John Meares, Esq., London, 1790, 2 vol., 4°; French edition, Paris, l'an III de la Republique (1794), 3 vol., 8°. German in G. Forster, I.

1789. Don Esteban Martinez's second voyage (cf. 1788). He anchored in Nootka Sound (Puerto de San Lorenzo des Perez, 1774; Cook's "Friendly Cove") on the fifth of May. Then he proceeded northward, but not above 50° N.

Lat. He sent Capt. Colnett as prisoner from Nootka Sound by way of San Blas to Mexico. It was not until October 28, 1791, that the English received Nootka Sound back through a treaty with Spain.

1789–90. Lieutenant George Mortimer's *Observations* on his voyage with the brigantine *Mercury*, under the command of Mr. John Henry Cox, to the NW coast of America [London, 1791] (cf. G. Forster, III, pp. 167–224; Sauer in Billings's Exped., pp. 182, 249).

This ship had been sent by the Swedish government to disturb the Russian trade in furs. Cox met Pribylov in Unalaska in 1790, but did not encounter the Billings expedition.

1789–92. First expedition of the Americans from Boston to the NW coast of America under Capt. Robert Gray in the sloops *Washington* and *Columbia*. On the seventh of May, 1792, he discovers the mouth of the Columbia River, which he named after his sloop (cf. G. Forster, I, p. 60; Fleurieu, in Marchand's *Voy.*, I, pp. ix, and cxi–cxv, IV, pl. II, after Meares, p. 219, where the course of the voyage is charted on a map; Vancouver's *Voyage*, I, p. 473; *Nouvelles annales des voyages* par Eyriès et Malte-Brun, Vol. X, Paris, 1821, p. 7).

1789, 1792, and 1793. Alexander Mackenzie's journey by land and river from Fort Chipewayan to 69°14′ N. Lat. (Mackenzie River), and his second journey from Fort Chipewayan to the Pacific Ocean, to the estuary of the Salmon River. He indicates erroneously that the mouth of the Columbia River is the same as that of the Tacoutche Tesse (cf. Fleurieu, in Marchand's *Voy.*, IV, pl. I). [Mackenzie's] *Voyages from Montreal Through the Continent of North America to the Frozen and Pacific Oceans in the Years 1789 and 1793*, London, 1801, 4°, pp. 121–412, etc., with maps; French edition, by J. Castera, Paris, 1802, 3 vol., 8°).

1789–91. Don Alessandro Malaspina's circumnaviagtion of the earth, and his charting of the coast from the mouth of the Rio de la Plata to Prince William's Sound (Bay Chugatsk). Cf. *Memorias sobre las observaciones astronomicas hechas por las navigantes Espanoles en distintos lugares del globo, los quales han servido de fundamento para la formacion de las cartas de marear publicadas por la direccion de trabajos hydrograficos de Madrid: ordenadas por Don Josef Espinosa y Tello, Gefe de Esquadra de la Real Armada, y primer Director de dicho estableciemento. De Orden superior. Madrid en la imprementa Real. Anno de 1802.* 2 vol. in 4°. Volume 2 here is of special interest, translated into French, with additions, by Wallenstein. In manuscript at the Hydrographic Dept. at the Naval Ministry in St. Petersburg; cf., *Zapiski Admiralteiskago Dept.* [Notes of the Admiralty Dept.] pts. III–XIII: "Puteshestvie Kap. Malespini v iuzhnoe more. Perevod s ispanskago"

[Voyage of Capt. Malaspina to the South Sea. Translation from Spanish]. Humboldt, *Nouv. Esp.*, I, pp. 338–340.

Malaspina, Espinosa, Don Josef Bustamiento y Guerro, Don Bauza, Murphy, and the botanists Thaddaus Hänke and Née, set sail from Cadiz on the thirtieth of July, 1789, with the corvettes *La Descubierta*, and *La Atrevida*. On the second of February, 1791, they arrive in Acapulco. There they receive the order to trace the Strait of Maldonado (Anian). They left Acapulco on the first of May, 1791. Three months later they reached Cape St. Bartolomeo (Quadra 1775, Cook 1778, Dixon 1706 [*sic*]); and the coasts from Cape Edgecumbe (Cabo Enganno, 57°01'30") to Montague Island. The location and elevation of Mt. St. Elias were exactly determined, the latter at 5,441 meters—2,793 *Tois.*, while La Pérouse had found 1,980 *Tois.* The Cerro Buen Tiempo [Mt. Fairweather] measures 4,489 m. Malaspina sought in vain for Maldonado's Strait; and he remained for a time in Port Mulgrave of Bering's Bay (59°34'20' N. Lat.). Then he turned back and reached Nootka Sound on August 13. In October of 1791 the *Atrevida* arrives at Acapulco, and the *Descubierta* arrives at San Blas. The locations of Sitka, Monterey, Guadalupe Island, and Cape San Lucas were astronomically determined on this voyage.

1790–92. Don Francisco Elisa and Don Salvador Fidalgo undertake an inspection of the west coast of North America with three ships. Cook's Inlet, Chugach Bay, under 60°54' N. Lat. Volcanic phenomena. Humboldt, *Nouv. Esp.*, I, p. 338.

1790–92. Vancouver's and Broughton's voyage. Proof is produced concerning the nonexistence of a passage to the East below 62° N. Lat., and the coast is surveyed from 30° to 61°30' N. Lat. Vancouver's achievements are universally respected and recognized, and his map must be recognized as the basis for all later ones. Cf.: *A Voyage of Discovery to the North Pacific Ocean and Round the World,...Performed in the Years 1790–92 [1795]...under...Capt. George Vancouver.* London, 1798, 3 or 6 vol., 4°; d° by William Robert Broughton. London, 1804, 2 vol., 4° with Atlas. French edition, 3 vol., 4° avec Atlas, Paris, l'an VIII (1799) and 2 vol., 8°, 1807.

1790–92. *Voyage autour du monde, pendant les années 1790, 1791, et 1792 par Étienne Marchand, précédé d'une introduction historique;...avec cartes et figures,* par C. P. Claret Fleurieu. Paris, l'an VI–VII, 4 vol., gr. 4°. Norfolk Bay (la Baya de Guadaloupe), Queen Charlotte Island, Nootka Sound. [Section on] botany, zoology, ethnography, and linguistics thoroughly edited. The geognostic material is without importance. The introduction by Fleurieu for the history of the voyage is very important.

1790. The Yakutsk resident Lebedev-Lastochkin and Company make ready the *St. George*, and in

1791 the *St. Paul*, commanded by Stepan Zaikov (cf. 1783–89).

His assignment was to enter Kenai Sound. Further reportings are wanting with regard to either of the two voyages (Berkh, pp. 115–119).

1791. In autumn, the geodesist Chudakov [Khudiakov?] was sent by Capt. Sarychev from Unalaska to Unimak, where he remained during the winter. He charted the first continuous, if imperfect, map of Unimak Island (cf. Lütke, *p. n.*, p. 291; Krusenstern, *Mem. Hydrogr.*).

1792. The voyage of the American Captain Roberts (cf. *Voyage dans les Etats-Unis d'Amerique, fait en 1795–97* par la Rochefoucauld-Liancourt, Paris, Du Pont, l'an VII [1798], 8 vol., in 8°, Vol. III, pp. 19–22; Fleurieu in Marchand's *Voy.*, I, pp. 594–606).

1792. Don Dionisio Galiano, Don Cayetano Valdez, Salamanca, and Vernace leave Acapulco on the eighth of March, as a result of suggestions made by Malaspina, the Viceroy, and by Comte de Revillagigedo, on the *goelettes* (galliotes) *Sutil* and *Mexicana*. During four months they survey the entire coast of Quadra Island. In George's Sound, or Canale de Rosario, they meet Vancouver and Broughton. And on their return voyage from Nootka to Monterey, they put in at Ascension Bay (Entrada de Heceta), which had been discovered by Don Bruno Heceta on the seventeenth of August, 1775. And Gray had discerned it in 1792 as the mouth of the Columbia River (cf. Humboldt, *Nouv. Esp.*, I, p. 340; and *Memorias sobre las observaciones astron.*, etc.).

1792. Don Jacinto Caamaño (and Torres?) goes to sea with the frigatte *Aranʒaʒu*, on the twentieth of March, by order of Viceroy Revillagigedo, in order to find Fuente's or Fonte's strait between 51° and 56° N. Lat. He surveys the north coast of Charlotte Island, the south coast of Prince of Wales Island (Prince de Galles), Revillagigedo Island, Bank Island or Calamidad Island, the Aristizabal Islands, as well as Pitt's Archipelago, which lies opposite Moñino Inlet. After a five-month voyage he returns again (cf. Humboldt, *Nouv. Esp.*, I, p. 343).

1792–97 and 1798–1803. The last privately owned Siberian ship, *Zosima and Savvatii*, outfitted by the merchant Kiselev from Irkutsk, makes two voyages. On the second voyage an island was sighted between 43° and 44° N. Lat., and 160° and 165° W. Long. from Greenwich. It was possibly in the process of rising over a relatively short period of time (Berkh, pp. 119–128).

During the years 1785–1798, the East India Company acquired a part of the fur trade with China, and sent at least one ship annually to the coast of North America (cf. Shemelin's *Journal der ersten Reise der Russen um die Welt von*

1803–1806 [Journal of the first voyage of the Russians around the world, 1803–1806]. St. Petersburg, 1816, 2 vol., 4°, Vol. II, pp. 332–334). But then the Americans, too, began to participate vigorously in these trade endeavors. From 1798 to 1802 (cf. Shemelin, l.c.) seventeen American ships touched the region of the Russian colonies. The fact that non-Russian trade endeavors were getting out of hand might have given Grigorii Shelikhov the major impetus for establishing the Russian American Company in the year 1799. Cf.: *Pod Vysochaishym Ego Imp. Vysochestva pokrovitel'stvom Rossiiskoi Amerikanskoi Kompanii, glavnago pravleniia akt i Vysochaishe darovanniia onoi kompanii pravila s priobshcheniem prilichnykh k onomu izakonenii* [Act of the board of directors of the Russian American Company under the highest protection of His Imperial Majesty and the rules imperially granted to that company with the addition of...]. St. P., pri Akad. Nauk., 1812, 58 pp., and the map in Russian [Cyrillic] characters published simultaneously for the region between 40° and 72° N. Lat., and 125° W. Long. and 136° E. Long.; the same in Storch's *Russland unter Alex. I,* vol. 1, pp. 145–162 and 265–297. Privileges granted to this company made further trade endeavors of individual Russian private companies impossible. Shelikhov's rival, Lebedev-Lastochkin, was hurt most of all by this situation. He had founded a settlement on Chugatsk Bay and had intended to expand far beyond. The leader of his work-troop (artel'), Vasilii Ivanovich [Vasilii Ivanov], had already undertaken a journey from Lake Iliamna into the country's interior. On that trip he traversed 500 versts. He met ten tribes with different languages, but found no settlements with more than 200 inhabitants. He often mentions a river, Tunka (Nushagak or Kuskokwim?), which was 4 to 6 versts wide; and over 40 villages lie along its banks. But Grigorii Shelikhov's and Golikov's company had always been of greater importance and volume than Lastochkin's company. Cf. *General'naia karta, predstavliaiushchaia sposoby k umnozheniiu Rossiiskoi torgovli i moreplavaniia po Tikhomu ili Iuzhnomu okeanu, s prilezhashchimi zemliami, znatneishimi ostrovami, prodolzhaiushchimisia ot Severo-Amerikanskago s Azieiu proliva do ravnodenstvennoi linii, s pribavleniem k tomu vnov naidennykh Kykhtak (Kod'iak) Afagnak i prochikh ostrovov. Takhe s podrobnykh opisaniakh, o nakhodiashchikhsia tam narodakh, seleniiakh i proizvedeniiakh; uchinennykh na korabliakh Severo Vostochnoi Amerik. Kompanii, Kapitana Golikova s tovarishchami (s shturmanom Izmailovym)* [General map representing the means for increasing Russian trade and sailing the Pacific or Southern Ocean, with adjacent islands, the most notable islands, extending from the strait between North America and Asia to the equinoctal line, with the addition of newly discovered Kykhtak (Kodiak), Afognak, and other islands. Also with detailed descriptions of the peoples, settlements, and products found there; made aboard the ships of the Northeastern American Company of Captain Golikov and comrades (with navigator

Izmailov)], 1787. And as hardcover edition: *Spetsial'naia karta Kykhtaka i prochim ostrovam, s pokaʒaniiem Severo-Amerikanskago berega, ʒalivov, gavanei, rek, lesov, i v kakikh mestakh rossiiskiia i tamoshniia seleniia nakhodiatsia. Opisan-nye tam byvshim kompanionom nashim Grigoriem Shelekhovym* [Special chart of Kykhtak (Kodiak) and other islands, with indication of the North American coast, bays, harbors, rivers, forests, and in what places Russian and local settle-ments are located. Described there by our late partner Grigorii Shelekhov] (1 l. fol.). In 1784 Shelikhov had founded settlements on Kodiak and Cook's Inlet (four years prior to the establishment of the settlement at Nootka Sound by Capt. Meares; cf. Simpson, *Journey Round the World*, I, p. 271). He put the Greek, Evs-tratii Ivanovich Delarov, in charge over the settlements as his deputy. This posi-tion was thereafter given to the Kargopol' merchant A. A. Baranov. He, in turn, founded New Archangel on Edgecumbe Island (Sitka) after the termination of the private companies and the establishment of the headquarters of the great company in 1799 in St. Petersburg, as ordered by Catherine II. Later, on New Archangel was transferred to the present Baranof Island. In 1812 Fort Ross was founded by Baranov's deputy, Commercial Councilor I. A. Kuskov. And in Sep-tember, 1821, the company received a new charter (*Reglement*).

1802–4. The voyage of the naval officers Khvostov and Davydov is of little inter-est with regard to natural history and geography. Cf.: *Dvukratnoe puteshestvie v Ameriku morskikh ofitserov Khvostova i Davydova* [Two voyages to America of the naval officers Khvostov and Davydov]. 2 pt., 8°, St. P., 1810–12. *Reise der Russisch-Kaiserlichen Flott-officiere Ch[wostow] und D[awydow] von St. Petersburg durch Sibirien nach Amerika und ʒurück in den Jahren 1802–1804* [Voyage of the Russian Imperial Naval Officers Khvostov and Davydov from St. Petersburg through Siberia to America and back in the years 1802–1804]; narrated by Gavril Ivanovich Davydov, and translated from Russian (into German) by Carl Johann Schultz. Berlin, 1816, 1 vol., 8°.

As early as 1799, Lieutenant Commander A. J. von Krusenstern had sub-mitted to the Russian government a detailed suggestion: to supply the American colonies with their necessities by the sea route from the Baltic Sea. This sugges-tion, however, found no interest for quite some time, until Naval Minister Mord-vinov and Chancellor Rumiantsev submitted it to Tsar Alexander I. Thereupon twenty-five expeditions were launched from Kronstadt into the South Sea and to the Russian American colonies between 1803 and 1827. Fifteen of these went at the cost of the government, nine at the Company's expense, and one was outfit-ted by Chancellor Rumiantsev. Cf.: Engelhardt's *Russian Miscellen.*, St. Peters-burg, 8°, Vol. I, pp. 28–69; *Severnyi arkhiv*, 1824, No. 11–18; Feier des 50-jährigen

Dienstzeit des Vice-Admirals v. Krusenstern [Celebration of the 50th Anniversary of the Service of Vice-Admiral von Krusenstern], by K.E. von Baer, *St. Petersburg. Zeitung*, 1839, No. 28–37.

1803–1806. Von Krusenstern's and Lisianskii's voyage around the world with the Extraordinary Ambassador to Japan, Privy Council Nikolai Petrovich Rezanov, von Langsdorff, Commerce Commissary of the Russian American Company Shemelin, botanist Dr. Tilesius from Leipzig, astronomer Horner from Zurich, Dr. Espenberg, and surgeon Leland. Cf.: *Reise um die Welt in den Jahren 1803–6 auf Befehl Seiner Kaiserl. Majestät Alexander I, auf den Schiffen...Nadeshda und Newa* [Voyage around the World in the years 1803–6, by order of His Imperial Majesty, Alexander I, on the ships...*Nadeẑhda* and *Neva*]. St. Petersburg, 3 vol., 4°, 1810–1812, with Atlas of 104 plates. Russian: St. P., v Morskoi tipografii, 1809–13, 2 vol., 4°; French: par J.B. Eyriés, Paris, 1821, 2 vol., 8°; Swedish: Oerebro, 1809, 2 vol., 8°; Dutch: Harlem, 1811 and 1815, 2 vol., 8°; Danish: Copenhagen, 3 vol., 12°. The following are closely related to the travelogue:

1) *Beyträge ẑur Hydrographie der grössern Oẑeane, als Erläuterungen ẑu einer Charte des ganẑen Erdkreises nach Mercator's Projection* [Contributions to the Hydrography of the larger Oceans, as commentary on a map of the entire earth according to Mercator's projection]. Leipzig, 1819, 4°, with map.

2) *Atlas de l'océan Pacifique.* St. Petersbourg, 1824 et 1827, 2 vol., fol. Vol. I, 1824, *Hémisphère austral*, 15 pl. Vol. II, 1827, *Hémisphère boréal*, 19 pl. Russian edition, St. P., 1823–1826.

3) *Recueil de mémoires hydrographiques, pour servir d'analyse et d'explication à l'atlas de l'océan Pacifique*, 2 vol., St. P., 1824–1827, 4°. Russian edition, 1823–1826, 4°.

4) *Supplémens au recueil de mémoires hydrogr., publies en 1826 et 27, pour servir d'analyse et d'explication à l'atlas de l'océan Pacifique.* St. P., 1835, 4°. Russian edition: St. P., 1836.

Langsdorff, who left the expedition in 1805, returning home on the land route through Siberia, published his *Bemerkungen auf einer Reise um die Welt in den Jahren 1803–1807* [Notes on a Voyage Around the World in the Years 1803–1807]. Frankfurt am Mayn, 1812, 2 vol., 4°, with copper engravings.

Lisianskii published his voyage under the following title: *Puteshestvie vokrug sveta v 1803, 4, 5, i 1806 godakh, po poveleniiu E. I. V. Aleksandra 1-go, na korable Neve, pod nachal'stvom flota kap.-leitenanta...Iuriia Lisianskago* [Voyage around the world in 1803, 4, 5 and 1806 by order of His Imperial Majesty Alexander I, on the ship *Neva*, under the command of Fleet Captain-

Lieutenant…Iurii Lisianskii]. St. P., 1812, 2 vol., 8°. The same in English: London, 1814, 1 vol., 4°.

Shemelin's travelogue and trading commentaries are found in: *Zhurnal pervago puteshestviia rossiian vokrug ʒemnago shara…Rossiisko-Amerikanskoi Kompanii glavnym kommissionerom Moskovskim kuptsom Fedorom Shemelinym* [Journal of the first voyage of Russians around the globe…by the chief commissioner of the Russian American Company, Moscow merchant Fedor Shemelin]. St. P., 1816–1818, 2 pts., 4°.

1804–06. The overland journey of the two American officers, Lewis and Clark. Cf.: *Travels to the Source of the Missouri River and across the American Continent to the Pacific Ocean. Performed by order of the Government of the United States in the years 1804–1806 by* Meriwether Lewis and William Clark. London, 1814, 4°, with maps.

At the mouth of the Columbia River they established a fort in 1815. The work mentioned, and especially the map, has served as the basis for most of the later journeys of discovery in the interior of North America. As early as 1806 there appeared in New York: *Message from the President of the United States, communicating discoveries made in exploring the Missouri, Red River, and Washita, by Capts. Lewis and Clark, Dr. Sibley, and Mr. Dunbar; with a statistical account of the countries adjacent. Read in Congress, February 19, 1806.* And in 1809 a volume in 8° was published in London, entitled: *The Travels of Capts. Lewis and Clarke,…from St. Louis, by way of the Missouri and Columbia Rivers, to the Pacifick [sic] Ocean; performed in the year 1804–06, by Order of the Govern. of the United States; containing Delineations of the Manners, Customs, Religion, etc., of the Indians, compiled from various authentic Sources and Original Documents, and a Summary of the Statistical View of the Indian Nations, from the official Communication of Meriwether Lewis.*

1805–12. Eleven promyshlenniks, deposited on the Komandorskie Islands by helmsman Potanov, by order of the [Russian] American Company, for the purpose of fur hunting, remain there (Golovnin, I, p. 161).

1806–07. Lieutenant Hagemeister sails to Okhotsk with the ship *Neva* of the Russian American Company. His ship remains there, while he returns home by land.

1807. Hedenström and Sannikov are ordered by Chancellor Rumiantsev to visit the Lakhof Islands in the Arctic Sea. They discover New Siberia.

1807–09 and 1811–14. Golovnin and Rikord sail on the sloop-of-war *Diana* from Kronstadt to Peter-Paul's Harbor and to Japan. Cf.: *Zapiski flota kapitana Golovnina, o prikliucheniiakh ego v plenu u Iapontsev v 1811, 12 i 13 godakh* [Notes of Fleet Captain Golovnin about his captivity among the Japanese in 1811, 12 and 13] (St. P., 1816, 3 vol., 4°); *Sokrashchennyia ʒapiski flota*

Kapitan-Leitenanta...Golovnina, o plavanii ego na shliupe "Diane," dlia opisi Kuril'skikh ostrovov, v 1811 g. [Abridged notes of Fleet Captain-Lieutenant...Golovnin on his voyage aboard the sloop *Diana* for a survey of the Kuril Islands in 1811] (St. P., 1819, 1 vol., 4°); *Puteshestvie Ross. Imp. shliupa Diany, iⱬ Kronshtata v Kamchatku* [Voyage of the Russian Imperial sloop *Diana* from Kronstadt to Kamchatka] (St. P., 1819, 2 pts., 4°); *Zapiski Admiralteiskago Dept.* [Notes of the Admiralty Dept.], pt. III, pp. 271–307; and *Zapiski flota kapitana Rikorda o plavanii ego k Iaponskim beregam v 1812 & 1813 godakh* [Notes of Fleet Captain Rikord on his voyage to Japanese shores in 1812 and 1813] (St. P., 1816, 1 vol., 4°). *Voyage de M. Golownin, contenant le récit de sa captivité cheⱬ les Japonais, pendant les années 1812–1813, et ses observations sur l'Empire du Japon. Traduit sur la version Allemande par Eyriés et suivi de la relation du voyage de Mr. Ricord aux côtes du Japon en 1812 et 13.* Paris, 1818, 2 vol., 8°. (See 1817–19.)

1808. Timofei Tarakanov's shipwreck. On a voyage of the Russian American Company sloop *St. Nikolai*, under the command of helmsman Bulygin, on the NW coast of America, about 47° N. Lat., i.e., at Vancouver's Destruction Island (cf. *Severnyi arkhiv*, 1822, No. 21 and 22, and *St. Petersb. Zeitung*, 1822, X, pp. 22–52).

1809. Helmsman Vasil'ev I, surveys part of the west coast of Baranof Island and of several adjacent islands.

1810. Lieutenant Rikord (and helmsman Khlebnikov) survey Shelikof Strait (cf. 1807–14).

1810–12. Mr. Astor, director of the New York Company for the trade in furs in the Pacific Ocean, readies the *Tonquin* under Capt. Jonathan Thorn for the Columbia River estuary. Cf. *Nouv. Ann. des voy.* par Malte-Brun, Vol. X, 1821, pp. 12–31. *Voyage par mer de New-York à l'embouchure de la Columbia,* based on English manuscripts, and published by Lapie.

1811–12. Voyage of Mr. Hunt and his companions from St. Louis to the mouth of the Columbia River, by a route through the Rocky Mountains. *Nouv. An.* par Malte-Brun, Vol. X, pp. 31–88.

1812–13. Voyage from the mouth of the Columbia River to St. Louis, by way of the Mississippi, by R. Stuart, with a map of the western part of the United States (cf. *Nouv. Ann. des voy.*, Vol. XII, 1821, pp. 21–113).

1813–16. Lieutenant Michael Petrovich Lazarev sails in October of 1813 on the ship *Suvorov*, of the Russian American Company, to New Holland and the colonies. On the fifteenth of July 1816, he returns to the harbor of Kronstadt (*Zapiski Admiralteiskago Dept.* IV, p. 419).

1815. Lt. Pavalishin's travels through the American colonies (cf. Engelhardt's *Miscellen.*).

1815–1818. O. von Kotzebue's first voyage around the world, with Shishmarev, Zakharin, Chamisso, Eschscholtz, Wormskjold, and Choris. Cf.: *Entdeckungs-Reise in die Süd-See und nach der Berings-Strasse, zur Erforschung einer nordöstlichen Durchfahrt, unternommen in den Jahren 1815–18,…auf dem Schiffe Rurick, unter dem Befehle des Lieutenants…Otto von Kotzebue* [Voyage of discovery to the South Sea and Bering Strait, to investigate a northeastern passage, undertaken in the years 1815–18, on the ship *Rurik*, under the command of Lieutenant Otto von Kotzebue]. Weimar, 1821, 3 vol., 4°. Russian edition, St. P., 1821–1823, 4°. Furthermore, Choris: *Voyage pittoresque autour du monde, avec des portraits…et accomp. des descriptions par M. le Baron Cuvier, et M. A. de Chamisso, et d'observations sur les crânes humains par M. le Dr. Gall.* Paris, 1822, 1 vol. fol. *St. Petersb. Zeitung,* 1822, VIII, pp. 19–25, and IX, p. 345.

1816–19. Hagemeister leaves Kronstadt for the colonies on the third of September with the Russian American Company ship *Kutuzov*, and Ponafidin leaves on the fifth with the ship *Suvorov*. The latter returns on the eighteenth of October, 1818, the former on the sixth of September, 1819 (*Zapiski Admiralteiskago Dept.* IV, pp. 424, 426, 482).

1816–1819. Roquefeuil (Camille de), *Journal d'un voyage autour du monde, pendant les années 1816–1819,* 2 vol., Paris, 1823, 8°. German edition: Jena, 1823, 8° (San Francisco, Nootka, Cape Flattery, Detroit de Fuca, Mt. St. Hyacinthe, Nouvelle-Archangel, Ile du Roi Georges, Kodiak, Ile du Prince de Galles, Nouvelle-Archangel, Sitka, Cape Chirikof, Christian's Sound, Frederick Sound, Detroit de Chatham, Ile de l'Amirauté, Cross Sound, Hood Bay, Detroit de Pitt, Nootka, Cape Mendocino, San Francisco, Sitka). Mainly trade information, but also nautical reports.

1817–1819. Golovnin's second voyage on the frigate *Kamtchatka* with Lütke, Wrangell, and Etolin (cf. *Puteshestvie vokrug suieta, Sovershennue na voennon shliupie Kamchatkie v 1817, 18 i 19 godak, flota Kapianom Golovnin* [Voyage around the world on the sloop of war *Kamchatka* in 1817, 18, and 19 by Capt. Golovnin] (St. P., 1822, 2 vol., 4°). (Petropavlousk, Bering, and Copper Islands, Near Islands, Kodiak, New Archangel, Fort Ross, Monterey, Bodega, California.) Map of Rumanzof Bay, Sitka Bay, and Chiniak Bay on the Kodiak coast.

1818–1819. The Russian naval cadet, Ustinyev [Ustiugov?], and the Company servants Kalmakov and Karssanovsky [Korsakovskii] survey the coast from the Shelekhov Sea [Shelikof Strait] to Cape Newenham (Lütke, *Voy. autour du monde; partie nautique,* p. 255).

1818. With this year there begin the relentless attempts of the English to reach the Pacific Ocean from the NE coast of America. John Ross, William Edward

Parry, John Franklin, Dr. Richardson, Buchan, Hood, Back, etc. These attempts, however, do not gain importance for those polar coasts closer to the Bering Sea, until the expedition of 1825–27.

1819–22. Voyage of discovery of the Imperial Russian sloop-of-war *Otkrytie* (Discovery), and *Blagonamerennyi* (The Well-Minded, Benevolent), under Lieutenant Commander Vasil'ev and Shishmarev, and the sloops *Vostok* (East) and *Mirnyi* (The Quiet One), under Bellingshausen and Lazarev. The first two vessels sailed for the Russian American colonies and parted company in 1820. Vasel'ev sailed the *Otkrytie* to Petropavlovsk. Shishmarev, with Dr. Stein and the astronomer Tarkhanov on board, sailed the other ship to Captain's Harbor on Unalaska, then to the islands Amchitka, Semisopochnoi, Goreloi, and then to the north side of the Andreanof and Fox Islands. On June 20, 1820, they visited St. J. Bogoslof Island. Icy Cape was reached, and they touched on Lawrence Bay and Nunivak (cf. *Zapiski Admiralteiskago Dept.* V, pp. 219–227, and *Khron. istoriia Aleut. ostrovov* II, pp. 1–20).

1819–21. Bellingshausen's and Lazarev's voyage took another course. They discovered a new island, Peter I, on the meridian of Cape Horn, and 15° west of there Alexander I. Cf.: *Dvukratnyia izyskaniia v Iuzhnom Ledovitom okeane i plavanie vokrug sveta v prodolzhenie 1819, 20, i 21-go godov, sovershennyia na shliupakh Vostok i Mirnyi, pod nachal'stvom Kapitana Bellingsgauzena* [Two investigations in the Antarctic Ocean and voyage around the world in 1819, 1820, and 1821 aboard the sloops *Vostok* and *Mirnyi* under the command of Captain Bellingsgauzen], with Atlas (St. P., 1831, 2 vol., 4°); *Zapiski Admiralteiskago Dept.* V, 1823, pp. 201–219; *Zapiski, ucheniia komandira Morskago Shtaba*, pt. XI, 1833, map.

1819–21. Ponafidin sails to the Russian American colonies with the Company ship *Borodino* (*Zapiski Admir. Dept.* IV, p. 482). The ship returns to the harbor of Kronstadt on 17 September 1821.

1819. Klimovskii's voyage along the Copper River (cf. von Baer's and Helmersen's *Beitr.* I, p. 161, and Khlebnikov's notes, ibid.).

1820–22. Dokhturov's voyage from Kronstadt to the colonies with the ship *Kutuzov*, of the Russian American Company.

1820–24. Wrangell's and Anjou's land and sea voyage to Cape Shalagskoi. With Wrangell: Midshipman Matiushkin, Dr. Kyber, and Helmsman Kosmin. With Anjou: Surgeon Figurin and Helmsman Ilgin. Part of the Asian polar coast was surveyed, to the Bering Sea. Cf.: *Physikalische Beobachtungen des Capt.-Lieut. Baron v. Wrangel, während seiner Reisen auf dem Eismeere...von G. F. Parrot* [Physical Observations of Lieutenant Commander, Baron Wrangell, during his voyage on the Arctic Sea, by G. F. Parrot] (Berlin, 1827); *Zapiski Admir. Dept.* V, pp. 259–328, VI, pp. 81–120, VIII, pp. 129–143,

XIII, pp. 179–217; *Zapiski ucheniia komandira Morskago Shtaba* I, 1828, pp. 144–149; *Puteshestvie po severnym beregam Sibiri i po Ledovitomu moriu, sovershennoe v 1820–24 godakh ekspeditsieiu, sostoiavsheiu pod nachal'stvom flota Lieut. Ferdin. fon-Vrangelia* [Voyage along the northern shores of Siberia and the Arctic Sea in 1820–24 by an expedition under the command of Fleet Lieutenant Ferdinand von Wrangell] (St. P., 1841, 2 vol., 8°).

1821–24. The voyage of the sloop *Apollo* [*Apollon*] and of the brig *Ajax* [*Aiaks*] under Capt. First Rank Tulub'ev and Capt. Second Rank Filatov. They left Kronstadt in September of 1821, in order to tack along the NW coast of America. Lieutenant Khrushchev assumed command of the sloop *Apollo* following Tulub'ev's death. The *Ajax* suffered shipwreck in 1821 (cf. *Zapiski Admir. Dept.* X, pp. 200–272: Plavanie shliupa Apollona v 1821–24 godakh [Voyage of the sloop *Apollon* in 1821–24]).

1821. The brigs *Riurik* and *Elisabeth* [*Elisaveta*], under Klochkov and Kislakovskii, leave for the colonies on the fifth of September. The *Elisabeth*, loaded with grain, was sold at the Cape of Good Hope.

1821 and 1822. Khromchenko's, Etolin's, and Vasil'ev's voyages of discovery with the ships *Golovnin* and *Baranov*. Cf.: *St. Petersb. Zeitung*, 1822, Vol. VIII, pp. 171–175, and *Russkii invalid* 1822, No. 266; furthermore, *Severnyi arkhiv*, 1824, No. 11-18, translated into German in *Berghausen, Hertha* II, pp. 190–222; 258–273; 583–604: "Bruchstuecke aus dem Reisejournal des Herrn Chromtschenko, gefuehrt waehrend einer Fahrt laengs den Kuesten der Russ. Niederlassungen in NW-Amerika, durch das sogenannte Seeotter-Meer im Jahre 1822" [Parts of the travelogue of Mr. Khromchenko, logged during a voyage along the coast of the Russian settlements in NW America, through the so-called Sea Otter Sea, in the year 1822]. Krusenstern, *Mem. Hydr.*, 1827, p. 108, and 1835, p. 98. They surveyed the coast between Norton Sound and Bristol Bay; and they discovered Nunivak.

1821–23. Schabelsky (Achelle): *Voyage aux Colonies Russes de l'Amerique...pendant les années 1821–23*. St. Petersburg, 1826, 8°, 106 p. Statistical remarks were published in *Balby's Geography*.

1822–24. Lieutenant Commander Andreas [Andrei] Petrovich Lazarev's and Capt. Second Rank M. P. Lazarev's sea voyage with the frigate *Kreiser*, and the corvette *Ladoga*. Brief report on the Russian American colonies and California in: *Plavanie vokrug sveta na shliupe Ladoge v 1822–24 godakh,...Kapitan-Leitenant Andrei Lazarev* [Voyage around the world aboard the sloop *Ladoga* in 1822–24,...Captain-Lieutenant Andrei Lazarev] (St. P., 1832, 1 vol., 8°); and *Zapiski Admir. Dept.* VI, 1824, p. 295, IX, pp. 457–467, XI, pp. 57–94.

1823–26. Otto von Kotzebue's second voyage around the world in the sloop *Predpriiatie*. Russian edition: St. P., 1828, 9 vol., 8°. German edition:

Weimar, 1830, 2 vol., 8°. English edition: London, 1830, 2 vol., 8°. Cf. also Hofmann (E.), *Geognostische Beobachtungen gesammelt auf einer Reise um die Welt* [Geognostic Observations, collected on a voyage around the world]. Berlin, 1829, 1 vol., 8°.

1823. Khromchenko surveys the Bay of Yakutat, especially Rurik Harbor.

1824. The sloop-of-war *Smirnyi* under Capt. Dokhtorov, dispatched to the colonies, turned back while still in the North Sea, and returns.

1824–26. Lieutenant Chistiakov and Murav'ev sail the Company ship *Helena* [*Elena*] from Kronstadt to the North American colonies.

1825–27. Wrangell with the war-transport-ship *Krotkii*, to Kamchatka, etc. Ship's logs in the Archive of the Hydrographic Dept. of the Naval Ministry at St. Petersburg.

1825–27. Franklin's second voyage from the Mackenzie River westward toward Bering Strait, while Richardson traveled up the Mackenzie River. Cf.: *Narrative of a Second Expedition to the Shores of the Polar Sea, in the Years 1825–27, by John Franklin, Captain..., including an account of the progress of a detachment to the eastward, by John Richardson.* London, 1828, 4°, 1 vol. with Appendix I: Topographical and geological notices by J. Richardson, pp. i–lviii.

1825–1828. F. W. Beechey is sent to meet Franklin from the Bering Strait side. *Narrative of a Voyage to the Pacific and Beering's Strait,...in His Majesty's Ship Blossom.* London: Murray, 1831, 2 vol., 4°. German: in the new *Bibliothek der Reisebeschreibungen*, published following the death of Bertuch (Weimar, 1832, Vol. LIX & LXI, 8°). Vol. I: Kamchatka, Lawrence Island, Shishmarev Entrance, Chamisso Island, Kotzebue Sound, Cape Franklin, Cape Lisburn, Icy Cape; Vol. II: Point Barrow, Kotzebue Sound, Bering Strait, St. Paul's Island, Aleutian Islands, Chamisso Island. Furthermore: *The Zoology of Capt. Beechey's Voyage.* London, 1839, 1 vol., 4°. Appendix, pp. 156–180: Geology by Buckland, Belcher, and Collie.

1826–29. Expedition under Capt. Fr. [Fedor] Lütke with the corvette *Seniavin*, with Baron Kittlitz, A. Postels, and Dr. Mertens on board; and with the sloop *Moller*, Capt. Staniukovich, with naturalist Kastal'skii and Dr. Isenbeck on board. Cf.: *Voyage autour du monde, exécuté par ordre de Sa Majesté l'Empereur Nicolas Ier, sur la corvette Le Seniavine, dans les années 1826–29 sous le commandement de Frédéric Lutké, capt.* Paris, 1835–36, 3 vol., 8°, et 1 vol., 4° avec *Atlas* in folio. Vol. I et II: *Partie historique*, Vol. III: *Les travaux de MM. les naturalists*, Vol. IV: *Partie nautique.* Russian edition: St. P., 1834–36, 3 pts. 8°, 1 pt. 4° s Atlasom. On Staniukovich cf. *Zapiski, ucheniia komandira Morskago Shtaba* III, 1829, pp. 103–125, the newly discovered Moller Island.

1828–30. Lieutenant Commander Hagemeister sails for the colonies with the Imperial Russian military cargo ship *Krotkii*, for the purpose of scientific investigations. Ship's logs in the Archives of the Hydrographic Dept. of the Naval Ministry at St. Petersburg. Cf.: *Zap. uchen. kom. Morsk. shtaba* III, 1829, pp. 182–191.

1828–30. Khromchenko sent to the colonies with the Russian American Company ship *Helena*. Ship's log in the Russian American Company Archive in St. Petersburg: *Zap. uchen. kom.* M. III. ch. IV, 1832, pp. 304–312.

1828–30. Erman (Adolph). *Reise um die Erde durch Nord-Asien und die beiden Oceane in den Jahren 1828, 29 und 30* [Voyage around the world through Northern Asia and the two oceans in the years 1828, 29, and 30]. Part one: Historical Report, Berlin, 1833–48, 3 vol., 8°, and map of Kamchatka, Berlin, 1838, 1 Bl. fol. by S. Schropp and Co. Vol. I: [Journey from Berlin to the Arctic Ocean in the year 1828], Berlin; Reimer, 1833; Vol. II: [Journey from Tobolsk to the Sea of Okhotsk, 1829], Berlin, 1838; Vol. III: [The coast of Okhotsk, the Sea of Okhotsk, and travels on Kamchatka in 1829], Berlin, 1848.

1829–33. Ross (John). Second voyage (cf. *Narrative of a Second Voyage in Search of a North-West Passage, etc.* London, 1835, 3 vol., 4°).

1829–30. Helmsman Vasil'ev's second summer-journey along the Kuskokwim River. Maps and diaries in the Archive of the [Russian] American Company at St. Petersburg.[144]

1829. Ingenstrom surveys the Andreanof Islands and some of the Near Islands. With him and later:

1830–32. Pilot Chernov Nuchek Harbor, Chtagaluk Island and the Kaknu [Kenai] River mouth.

1830–32. Vasil'ev II surveys the southern coast of Alaska in baidaras from Cape Douglas to Cape Kumlun. (Kukak Harbor and Wrangell.) Cf.: Lütke, Vol. IV, *partie nautique*, p. 274, and the map, published by the Hydrographic Dept. of the Naval Ministry, 1844. Berghaus, *Annal.* III, pp. 390–392.

1831–33. Capt. Khromchenko sails to the colonies with the Imperial Russian cargo ship *Amerika*. Ship's logs in the Hydrographic Dept. of the Naval Ministry.

1831. Lieutenant Teben'kov makes an exact survey of Norton Bay (Krusenstern, *Mem. Hydr.*, 1835, p. 98).

1832. F. von Wrangell's journey of inspection to the Nushagak River (Lütke, *p.n.*, p. 267).

1833–34. Capt. Etolin works at surveying the Prince of Wales Archipelago. Zarembo at Wrangell Island, etc.

1834–36. Capt. Schanz [Shants] sails to the colonies with the Imperial Russian cargo ship *Amerika*. Ship's log at the Archive of the Hydrographic Dept. of the Naval Ministry at St. Petersburg.

1834–37. Overland expedition of Midshipman Glazunov along the Kvikhpak [Yukon] River. His journal is in the Archive of the Russian American Company.

1835. Capt. Teben'kov sails to the colonies with the ship *Helena* [*Elena*], of the Russian American Company. The latter remains there.

1836–37. The Creole Kolmakov journeys with baidaras up the Kuskokwim River. His diary in the Archive of the Company.

1836–37. Helmsman Voronkovskii surveys the south coast of Alaska from Cape Khitkuk on, Unga, and the Pribilof Islands. Cf.: Lütke, *partie nautique*, and Baer and Helmersen, *Beitr.* I, p. 323. His maps and journals at the Russian American Company.

1836–39. *Narrative of the Discoveries on the North Coast of America, Effected by the Officers of the Hudson's Bay Company During the Years 1836–39*, by Th. Simpson and Dease. London: Bentley, 1837 [1843], 1 vol., 8°, with maps. The Hudson's Bay Company leases in 1838, for ten years, from the Russian American Company, certain areas inhabited by the Kolosh (Tlingits), for the purpose of the fur trade.

1836–39. Du Petit-Thouars: *Voyage autour du monde*. Paris, 1840–44, 4°. Touches only upon Kamchatka; therefore, of no importance here.

1836–42. Capt. Belcher (E.). *Narrative of a Voyage Round the World...During the Years 1836–42*. London, 1843, 2 vol., 8°. Vol. II, p. 331: the NW region of America. Vol I: Port Etches, Cape Hammond, Cape Suckling, St. Elias, Port Mulgrave, Norfolk Sound, Edgecumbe, Sitka, Kodiak.

1837–39. The ship's mate Malakhov journeys along the Kvikhpak River. His notes in the Archive of the Company.

1837–39. Capt. Berens sails to the colonies with the Company ship *Nikolai*. Dr. Fischer is the ship's surgeon.

1838. Helmsman Kashevarov's inspection and survey of the North American polar coast from Point Barrow eastward. The map and the journal of the voyage are in the Archive of the Hydrographic Dept. of the Naval Ministry at St. Petersburg, and that of the Russian American Company. Cf.: *Syn otechestva* 1840, No. 1, pp. 127–155;[145] *Sankt Peterburgskiia vedomosti* 1847, No. 190–193. Or Erman's *Archive*, Vol. V, 1847, pp. 389–390.

1838. Lindenberg charts Admiralty Island and Chilkat River on Lynn Canal.

1838–42. Wilkes (Charles). *Narrative of the United States Exploring Expedition During the Years 1838–42*. Philadelphia, 1844, 7 vol., 4° with atlas. Vol. IV: Cape Disappointment, Juan de Fuca Strait, Admiralty Inlet, Fort Nisqually, Mt. Rainier, Columbia River, Astoria, Fort Vancouver, etc.

1839–40. Helmsman Murashev charts Kupreanof Strait and the corresponding coasts of Kodiak and Afognak.

1839–41. The Company ship *Nikolai*, under Capt. Kadnikov, sails for the colonies. And on board the same, by order of the Academy of Sciences, is the preparator of the Zoological Museum, Il'ia Voznesenskii.* The ship *Nikolai* returns in 1841 under the command of Capt. Voevodskii, with Dr. Blaschke on board.

> *I. G. Voznesenskii leaves in August of 1839. In 1840 and 1841 he visits New Albion, upper and lower California; 1842 and 1843, the Aleutian Islands, several groups of islands in the Bering Sea, and Kotzebue Sound; 1844 the Kuril Islands; in 1845 and 1846 he journeys along the coast of Okhotsk; and in 1847 and 1848 throughout the entire Kamchatka Peninsula. And from here he returns to St. Petersburg at the end of July of 1849, with the ship *Atkha*, by way of Sitka (cf. 1847–49).

1840. On the fifteenth of August, D. F. Zarembo leaves for the colonies with the Company ship *Naslednik Aleksandr*. He reaches New Archangel on the third of April 1841.

1840–42. The Naval cargo ship *Abo*, under Lieutenant Commander Junker [Iunker], sails to Okhotsk and the American colonies (cf. *Zapiski Gidrograficheskago Dept. Morskago Ministerstva* II, 1844, pp. 164–223, article by A. I. Butakov).

1841–42. Simpson (G.), Governor in chief of the Hudson's Bay Company's territories in North America: *Narrative of a Journey Round the World During the Years 1841 and 1842*. London: Colburn, 1847, 2 vol., 8°. Sitka, Taku, Stikine, Fort Simpson, Grenville Canal, Fort McLaughlin, Red River Settlement, Fort Vancouver, Sitka, Sandwich Islands, Aleutian Islands, Okhotsk, Urals, St. Petersburg. *Ibidem*: Duflot de Mofras, voyage to San Francisco, Monterey, Santa Barbara; and Grenough to the Columbia. The Russian American Company assigned to Mr. Simpson Mr. N. von Freimann, whose manuscripts are in part at the Archive of the Company. Cf.: Izvlechenie iz otcheta G. Freimana, ezdivshago chrez vladeniia Gudzonbaiskoi Kompanii [Extract from an account by Mr. Freimann, traveling through the possessions of the Hudson's Bay Company]. Article by Mr. Savich in *Zapiski Russkago Geograficheskago Obshchestva* I, pp. 79–93, and Erman's *Archive*, VI, pp. 226–240.

1842–44. Lieutenant Zagoskin's voyage to the Kvikhpak [Yukon] and Kuskokwim Rivers, by order of the Russian American Company. Cf: *Peshekhodnaia opis' chasti russkikh vladenii v Amerike, leit. L. Zagoskinym* [Pedestrian survey of part of the Russian possessions in America, by Lt. L. Zagoskin]. St. P., 1847, 2 pts., 8°. *Zapiski Russkago Geograficheskago Obshchestva* [Notes of the Russian Geographical Society], St. P., 1846, pp. 135–202. Izvlechenie iz dnevnika

Zagoskina [Extract from Zagoskin's diary]; *Zapiski Gidrograficheskago Dept. Morskago Ministerstva* IV, 1846, pp. 86–102 (Norton Sound).[146]

1843 and 1844. Capt. Fremont (J.C.) *Report of the Exploring Expedition to the Rocky Mountains [in the Year 1842, and to] Oregon, and North California [in the Years 1843–44).* Washington, 1845, 1 vol., 8°.

1843–45. The Imperial Russian cargo ship *Irtysh*, goes to the colonies under Capt. Vonliarliarskii, and remains there. *Zapiski Gidrograficheskago Dept. Morskago Ministerstva* II, 1844, pp. 432–436; III, 1845, pp. 387–392.

1844. Skipper Malakhov inspects the Susitna River, by order of the Russian American Company. Simultaneously, Operations Manager (*Prikashchik*) Grigor'ev inspects the Copper River. The journals of both are in the Archive of the Company.

1846–48. The ship *Sitkha* is sent to the Russian American colonies under Skipper Konradi.

1847–49. The Company ship *Atkha*, under Skipper Ridell, makes the voyage to the colonies and back. On board is the mining engineer, Lieutenant Doroshin, who was ordered to go to the colonies at the request of the Russian American Company. Exploratory work on Baranof Island, California, and Cook's Inlet (Kaknu [Kenai River]).

1847–48. Helmsman Serebrennikov's inspection of the region between the sources of the Copper and Kvikhpak Rivers. He, as well as two other Russians and six Aleuts, was supposedly murdered at the sources of the Copper River.

1848. The Imperial Russian cargo ship *Baikal*, under Captian Nevel'skoi, sails for the colonies and remains there.

1848–49. The English captains Kellett and Murr [Moore] are sent to Bering Strait with the ships *Herald* and *Plover*, to look for Franklin. Cf.: *Morskoi sbornik* II, 1849, No. 7.

1848. The ship *Sitkha*, Capt. Konradi, sails for the colonies again and is supposed to return in 1850.

1849. The company ship *Atkha*, under Ridell, sails to the colonies for the second time.

Notes

1. In 1847 Helmsman Serebrennikov had been sent by the Russian American Company to investigate the region between the sources of the Copper River and the Kuskokwim. Near the springs of the first-named river, he, two other Russians, and four Aleuts were allegedly murdered by the Kolosh of the Tundra (Tundrskie K.).

2. This fate will certainly come unexpectedly early to the above-named geognostic remarks concerning California, inasmuch as they have been written and published worldwide prior to the discovery of the gold deposits in Upper California.

3. A beautiful picture of the same is in Wilkes's *Travels* [*Expl. Exped.*], Vol. V, p. 252. The sources concerning the location of Mt. Shasta diverge from one another. Wilkes moves it much closer to the coast than Fremont does (Kiepert). Wilkes also calls the Sierra Nevada the California Range (Vol. V, p. 159). Regarding this point, the maps from the travelogues of Wilkes, Fremont, Simpson, Hunt, Stuart (as published by Lapie), Lewis and Clark, etc., may be compared. Also Kiepert's map of Mexico, Texas, and California (Weimar, 1847), and the map of the Hydrographic Dept. of the Naval Ministry (St. Petersburg, 1848, No. 11, Monterey to Charlotte Sound). On the enclosed Map No. 1, a sketch made by Mr. Laframboise, then Artillery Lieutenant, as well as Löser's maps, etc., were used for the part of California north of the Bay of San Francisco.

4. Cf. the remarks concerning California in last year's issue of this journal, p. 151.

5. This river is the Tlamath (Clamet and Shasta [Klamath]) on English maps.

6. Cf. Transactions of the Geographic Society of St. Petersburg (*Zapiski Russkago Geograficheskago Obshchestva*), Vol. I: Remarks concerning a journey from Sitka through the possessions of the Hudson's Bay Company, by Freimann. Translated in part from Russian in A. Erman's *Archiv für wissenschaftliche kunde von Russland* [Archive for scientific information about Russia], Vol. VI, pp. 226–240. Essays and translations into other languages are extant. But such publications ought best not take place when translator and publisher are not familiar with the subject at hand, or if they do not want to make the effort to use the already existing material. This, for instance, has happened with the report given by Mr. Freimann, who apparently made no attempt to publish it either as a scientific achievement, or as research. He merely wanted to raise questions and make comments, without offering thorough explanations, which would have been possible in this case. The publisher of the *Archive* seems to have known neither Fremont's travelogue, nor the names Sutter (l.c., p. 237), or Capt. Wilkes (p. 238). Mr. Freimann did meet Capt. Wilkes, at Ft. Vancouver on the Columbia [River], 28 August 1841 (Wilkes, *Expl. Exped.*, Vol. V, p. 126).

7. According to Mr. Beck, plutonic and volcanic rock formations are much in evidence along both sides of the Columbia River (cf. A. Boué in *Bulletin de la Société géologique de France*, Vol. I, 2-ème série).

8. This mountain, also called Umpuquois, with its high, flat summit, was sighted by Cook (*Third Voyage*, Vol. II, p. 4) from Cape Fairweather (44°55′ N. Lat.) It was named by Capt. Clark in April of 1806 (Lewis and Clark, *Travels...1804–1806*, p. 502). Mt. Jefferson, as well as Mt. St. Helens, is perfectly cone-shaped.

9. Measured at 1,203 *Tois.*, according to Gardner's trigonometric calculations. In its vicinity is a burning mountain, perhaps Mt. Jefferson. The report that Mt. Hood has a supposed elevation of 16,500 ft. Engl. is therefore to be rejected (cf. also Berghaus, *Geography and Ethnography* II, p. 763).

10. This according to Vancouver, II, p. 243. According to Gardner it is volcanic. Capt. Clark (p. 502) noticed no volcanic activity on Mt. St. Helens, nor on Mt. Rainier (cf. also Simpson, *Journey Round the World*, Ch. III and IV).

11. The north coast of Charlotte Island (at the border strait or Casa de Cox, cf. Douglas, 1789, in Meares's *Voyages*, p. 365) is mostly steep, but not high; and it is strewn with rock boulders, according to E. Marchand (Vol. I, p. 522, and Vol. IV, Pl. IX). These might have been torn loose from adjacent rocks during earlier earth movements. These boulders consist of a conglomerate of flint, brown ironstone, and a grey, not very hard kind of stone (graywacke?). On the same latitude as Charlotte Island there is, near the mainland, Pitt's Archipelago. Grenville Canal, which belongs to it, is described by Simpson thus (I, p. 233): "We saw (on its course) some beautiful waterfalls, which had been greatly increased by the late heavy rains, tumbling down the sides of the mountains, where they found so little soil, that they carried their foam to the sea just as pure as they had received it from the clouds."

12. On the northeast side of Vancouver or Quadra Island, a river empties into the sea, on the banks of which layers of coal appear on the surface. The steam boats of the Hudson's Bay Company get coal here for a nominal price. In earlier times the coal was shipped on a six-month journey across the big Salt Lake to Fort McLoughlin (*Times*, 1 February 1848). Cook says of Nootka Sound (Vol. II, p. 33): "Besides the stone or rock that constitutes the mountains and shores, which sometimes contains pieces of very coarse *quartz*, we found, amongst the natives things made of a hard black *granite*, though not remarkably compact or fine grained"(cf. Vancouver's *Atlas*, Pl. XIV: Depiction of the sound). Roquefeuil (I, ch. IV and II, ch. XII) has given that sound and the island the most thorough treatment. On p. 210 he mentions a Mt. Tasche, located in ca. 50°N. Lat., 126°30′ W. Long.

13. In 1775, according to the diary of Don Antonio Maurelle (Appendix II) several active volcanos could be seen from Port Bucarelli on this island. One of them might have been Mt. Calder (56°15′ N. Lat., 133°30′ W. Long.). Vancouver did not see any smoke on Mt. Calder in 1793 (Vancouver, *Atlas*, Pl. XV, Profile).

14. From now on those regions are described, for which reports were of geognostic interest—regions from which records were extant from the collections of I. Voznesenskii, Kashevarov, Fischer, E. Hofmann, and Postels (cf. Appendix II, 1839, 1838, 1837, 1826 and 1823).

15. The same was first known by the name Baya de Guadalupa by the Spaniards in 1775 (Journal of Antonio Maurelle, in Pallas, *N. B.* I, p. 290, and Fleurieu in Marchand's *Voy.*, I, p. l). The English called it Norfolk Bay in 1787 (Dixon, p. 180). And thereafter Capt.

Prosper Channal called it Chinkitane (E. Marchand, Vol. I, ch. IV, pp. 209–288, and Vol. IV, pl. VIII). Vasil'ev I and Golovnin provided the first more detailed survey of Sitka Sound (Vol. I, ch. VI; Vol. II, ch. III). But not until 1848 was a special map published by the Hydrographic Department of the Naval Ministry at St. Petersburg. From this was copied the attached map.

16. Belcher's *Voyage*, Vol. I, p. 96, has one of the best views of Novo-Arkhangel'sk known to date.

17. Cf. Appendix II, 1823–1826.

18. Cf. Simpson, *Voyage Around the World, 1841–1842*, Vol. II, pp. 174–176.

19. Also mentioned by Erman are deposits of Tertiary brown coal on Sitka (*Reise um die Erde* [Voyage around the world], Vol. 3, p. 213).

20. I leave this report in place, although according to several inquiries of mine, even the oldest natives (Kolosh) of Sitka claimed to have no knowledge of these phenomena. Certain is that between 1775 and 1791 the volcanicity of Mt. Edgecumbe is mentioned neither by the Spaniards (Journal of Antonio Maurelle, Pallas, *N. B.* I, p. 270), nor by the English (Cook and Dixon, pp. 180–185), nor by the French (Marchand, Vol. I, ch. IV, Vol. IV, Pl. VIII). The present Mys Otmelyi Vneshny [Shoals Point] was called Cape Whites by Dixon. And the promontory between there and Cape Edgecumbe or Sitka was called Cape del Enganno by the Spaniards. Maurelle's diary (in Fleurieu, Marchand, Vol. I, p. lxix) states: *La montagne a pour base le Cap qui se projette au large: sa forme, est-il dit dans Journal, est la plus belle et la plus régulière qu'on ait jamais vue: elle est isolée et détachée de la chaîne des autres montagnes: son sommet étoit couvert de neige; on voyoit au-dessous quelques grands espaces nus qui se prolongeoient jusque vers le milieu de ses flancs; et de cette hauteur jusqu'au pied, sa surface étoit couverte de grand arbres.* From Lisianskii's *Puteshestvie* (Vol. II, p. 126) we translate: "Judging by the circumference and depth of the crater, one ought to believe that this volcano must once have been much higher, that gradually it ceased its activity completely, collapsing in on itself, whereby it filled up the abyss, which once had emitted flames. Since then probably a long time has elapsed, because some of the expulsions of the crater had already changed into the composition of fertile ground."

Remarks regarding Kruzof or Edgecumbe Island: Mr. Lieutenant Commander A. P. Solokov will shortly publish his edition of Chirikov's voyage. We owe him thanks for his kind notice, viz.: that Chirikov without doubt did sight Mt. Edgecumbe in 1741; but he noticed no volcanic phenomenon on the same. La Pérouse (II, p. 221, August 6, 1786) remarks about this region: *Le cap Enganno...est une terre basse couverte d'arbres, qui s'a-vance beaucoup dans la mer, et sur laquelle repose le mont Saint-Hyacinte, dont la forme est un cône tronqué, arrondi au sommet; son élévation doit être au moins de deux cents toises.*

21. A glance at the map suffices to leave no doubt in this regard, namely, that Sitka and Chichagof Islands, for example, had formerly been connected to each other and to the northern part of the mainland.

22. According to Cook (*Third Voyage*, [transl.] by G. Forster, p. 70), the SW point of Kaye Island is a high, naked rock. Everywhere the island falls off obliquely to the shore; at the foot of it there is a narrow beach, covered with rounded rocks. The cliffs consist of

bluish, in part reddish, weathered rock (clay-slate?), and the mountains are covered with schist up to half their height.

23. In Cook, l.c., p. 69, is stated: "The point of the Cape is low; but within it is a tolerably high hill, which is disjoined from the mountains by low land; so that, at a distance, the Cape looks like an island." The islands, which Cook locates in Controller Bay (west of Cape Suckling), are missing in all later reports.

24. Cook (Vol. II, p. 4) says that at 45° N. Lat. there are already such white cliffs that are probably glaciers. "[T]he bare grounds toward the coast were all covered with snow, which seemed to be of a considerable depth between the little hills and rising grounds; and, in several places toward the sea, might easily have been mistaken, at a distance, for white cliffs." (This was at the beginning of March 1778 in the same latitude as Bordeaux, Turin, the mouth of the Danube River, Stavropol.) The glaciers in Kotzebue Sound, however, in 161°42′20″ W. Long. and 66°13′25″ N. Lat. (mentioned in Gilbert's *Annales*, Vol. IX, 1821, pp. 143–146, also in Kotzebue's first voyage, 1815–1818, Part I, p. 146) have been proven by Beechey to be false (German ed. I, pp. 403–406).

25. This volcano has later been called Mt. Wrangell (cf. Map II). It is located at 62°N. Lat. and 142°–143° W. Long.

26. Mining Engineer Doroshin is said to have found gold soaps [?] (clay with a content of gold) on the Kaknu River. Report by letter and orally. [Note from p. 421, re: Chugach Bay]: At the northern coast of this bay, at 60°54′ N. Lat., Don Salvador Fidalgo witnessed an eruption in 1790. The natives led him onto a snow-covered plain, where large masses of ice and rocks were catapulted upward to astonishing heights (Humboldt, *Nouv. Esp.*, I, p. 338).

27. Mining engineer Lieutenant Doroshin has sent this coal, and especially that from Anchor Point (M. Iakornoi), as a sample to California, where it has been recognized as useful for use in steamships.

28. Lütke (*partie nautique*, p. 266) has this chain beginning at the Susitna as a continuation of the Yakutat Mountains.

29. This mountain is perhaps Bering's Mt. St. Dolmat, which was dormant in 1741 (neither Bering nor Steller speaks of it) (Pallas, *N. B.* I, p. 269; cf. also Cook's *Third Voyage*, II, p. 108, and G. Forster's *Geschichte der Reisen* [History of the travels], etc., II, p. 64, with a representation of the fire-belching mountain at the Cook River, according to Portlock and Brain [?], from 25 July 1786). Thus, this volcano was not first made known by Vancouver (as Wrangell indicates), but earlier than that, i.e., since 1778, by Cook. On August 1, 1779, Arteaga saw it smoking (cf. Appendix II, 1779).

30. Cape Kumlun [Kumliun], according to Lütke, at 56°32′02″ N. Lat. and 4°10′ west from Cape Douglas.

31. This is the commonly used name of the peninsula in the Russian American Colonies. The Russian *Aliaksa* sounds like the German Alaeksa. Hereinafter: Alaska, which in this work refers to the peninsula, not the mainland.

32. Close to that bay, the Dark-Brown (Chernoburyi) Island with Cook's Augustine Mountain.

33. I assume that such an investigation is permitted here, inasmuch as these regions are little known and have not been described since Lütke. And during most recent times there might have occurred alterations in the terrain, such as have happened on the Aleutian Islands. In Coxe, p. 254, in Pallas, *N. B.* I, p. 255, and Busching's *Mag.* XVI, p. 269, there, in Krenitsyn's Voyage, are found speculations concerning occurrences of alterations of the coast. "The galliote St. Catharina wintered near Alaska Island from 1768–1769. Although the instructions for Captain Krenitsyn indicated that a ship of a certain private owner had found a comfortable harbor there in the year 1762, people looked for it in vain. The entrance to the channel behind this island is very difficult on the NE side because of strong currents, the ebb and flow of the tide, and is shallow. But on the SE side the inlet is much more comfortable, because here it is up to 5.5 fathoms deep. Along this strait, and along the rest of the coast of Alaska, many volcano-like mountains were visible, but very little vegetation. Therefore, it is plausible that, since the year 1762, either significant alterations might have occurred along this coast, or the earlier reports about it might have been unfounded." Cf.: in the sequel the description of Bering Island, Umnak, and Unalaska, and Veniaminov, I, p. 7 (Alaska).

34. Iantarnyi Zaliv, named thus probably not without reason. The natives exchange amber with the people of Kodiak (cf. Kodiak). It follows that amber is found here.

35. Cf. Veniaminov, I, p. 223, and Chamisso in Kotzebue's voyage of 1815–1818, pp. 164–165: "The two peaks on Alaska Peninsula are extraordinarily high. The first in the northeast (most likely the Pavlovskii Volcano), which a few years ago (1786) collapsed in on itself after an outbreak, seems to be the higher one of the two, with its blunted summit. The next one (Medvenikovsky or Morshevskoi?), is an acutely pointed cone. It appears to be higher than the one on Unimak." [Addition from the errata sheet of the original edition, p. 421:] On Helmsman Iakov's map to Krenitsyn's and Levashev's voyage, a volcano on the Alaska Peninsula, perhaps the Pavlovskii Volcano, was indicated for the first time (1769). And in the narrative to this voyage, several volcanos were mentioned (Coxe, German, p. 192).

36. On the slopes of the Aleutian chain of volcanos, close to where they connect to the American Continent, i.e., on Alaska and Unga, there are deposits of black soft coal with contents of amber, according to Erman (I, 3, p. 212).

37. This might be the volcano mentioned by Sarychev (Vol. II, p. 31, 1790). In that place he states: "Opposite the island Unachokh, there is on Alaska a high, fire-spouting mountain, the summit of which had collapsed during an eruption in 1786 with a mighty boom." Compare, however, the preceding comments concerning Pavlovskii Volcano. Veniaminov does not mention this mountain.

38. On Berghaus's map (*Phys. Atlas*, section 5, Geology, No. 6) I find the erroneous indication of a volcano. Moreover, the names of Alaska's three volcanos are confused. The same is the case in Johnson's *Physical Atlas*. [Note from the errata sheet, p. 422 of the original text:] Krusenstern (*Mem. Hydrogr.*, 1827, p. 103) erroneously believes that conical mountain of Cook's to be on Alaska. Cook saw it smoking on the twentieth of June, 1780, and fixed its location at 64°48' N. Lat., and 195°45' E. Long. [164°15' W. Long.]. This indication, however, speaks for Mt. Shishaldin on Unimak. The most recent

maps of these regions carry only the Russian names of these islands. Therefore, it is difficult to determine Sarychev's islands. Nanimak is perhaps Chernoburyi or Gusinyi; Animak perhaps Olenii, Laliaskikh Island the Dolgoi or Ilashek Island. But Kikhdokh Island, Kitagotakh, and Unachokh must be clarified on location.

39. Cf. coastal profiles in Lütke's *partie nautique*: Vues de la côte N, de la presqu'ile d' Aliaksa No. 1–5, incl. 6 et 7.

40. Cf. Appendix II, 1826–1829, Lütke's expedition.

41. Those layers indicated south of Moller Bay and NE of Perenosnyi Bay, with petrifications and hot springs in them, are either of the same substance, or the latter belong to the same formation (cf. Unga).

42. Because of a lack of space, these geographic conditions could not be exactly indicated on the accompanying map. It is extremely difficult, generally, to prepare a map which indicates the geologic conditions and mountain elevations together. It would have been avoided here, had not other reasons made this necessary.

43. On our maps we have called this point Veniaminof Volcano. [Margin note in pencil in the Rasmuson Library copy: Black Peak. Not! Volcano in 1892.]

44. Cf., Appendix II, 1851-1852.

45. Cf. *The Zoology of Captain Beechey's Voyage*, London,1839, 4°, Appendix, pp. 165–180: Geology by Buckland, Belcher and Collie.

46. Governor of the Russian American Colonies, 1835–1840.

47. Cf. Appendix II, 1826–1829, Lütke's expedition.

48. Cf. P. Erman's "Contribution to the Monography of the Marekanite [a low-water-content obsidian with rounded surfaces]," in *Treatises of the Berlin Academy*, 1830, and Erman's *Reise um die Erde*, I, Vol. 3, pp. 94–96. Here, too, the proximity of the graywacke formations seems to substantiate that the marekanite rock, and those half-molten rocks found with it, resulted from the subsequent influence of plutonic masses on layers of graywacke. On the whole the geognostic conditions of the region resemble those of the Marekanian mountain range.

49. Vivianite and different kinds of clay also occur in the western half of Kamchatka (Erman, I, 3, p. 165).

50. According to Lütke, *p.n.*, p. 197, "à crête arrondie."

51. According to Collie, a 2.5-mile-long wall, up to 90 ft. high, which contains fossilized bones in its eastern portion.

52. Cf. the map of the Hydrographic Department of the Naval Ministry, 1849. Kupreanof Strait runs southeasterly into Karlukskii and Malinovoi Straits, which surround Malinovoi or Severnoi Island [Raspberry], and Little Malinovoi or Severnoi Strait with Kitoi Island or Govorushchi Island [Whale Island] between them.

53. To date only the Chernichef Mountains north of Pavlof Harbor have become known to us.

54. It can nevertheless not be denied that, judging by the presence of the sandstone and that of clay-slate, soapstone-shale, and foliated chlorite, chalcedony, jasper, carnelian and brownstone, it could easily be believed that in this place there should exist the rare preconditions for the formation of diamonds by the wet process. This would take place by a

peculiar deterioration process of organic substances (mutual influence of ferric oxide and decaying organic substances).

55. Sarychev (II, p. 29), by contrast, claims: "At the NE end of the island there is a mountain in a tripartite shape."

56. Thus the verbatim translation of *neskol'ko peresheikov*. The meaning of this passage is apparently this: that several bridges crossed that glowing fissure; or at least, so it seemed, as seen from afar off. From Baer.

57. [Note from the original errata sheet, p. 422:] The Krenitzin Islands were named by Krusenstern (*Mem. Hydr.* 1827, p. 94). But he does not count Unalga, Akutan, and Akun in this group. Krenitsyn was the first to see them (1768–1769). Later, Solov'ev saw and described them (1770–1775).

58. [Cf. also C. H. Merck, *Siberia and Northwestern America*, p. 68].

59. [C. H. Merck, too, mentions this phenomenon. Cf. *Siberia...*, p. 67].

60. Thus, among the natives there is preserved the legend of a great earthquake, during which, they say, the mountains struggled with each other; and that Makushin on Unalaska remained the victor.

61. The snow-limit of Unalaska ($53°44'$–$52'$ N. Lat.) varies greatly according to different reports. According to Lütke, 556 *Tois.* or 1,083 meters, according to Chamisso, 1,657 meters, or 5,435 ft. median elevation. The number is certainly too high. But it comes close to Erman's measurement of Shiveluch Mountain (on Kamchatka, at $56°40'$ N. Lat.), which has a snow-limit of 822 *Tois.* or 1,605 meters. According to Veniaminov, I, p. 4, snow recedes higher upward in summertime on the coastal mountains, than it does on mountains inland. In fall, the situation is reversed.

62. Captain's Bay or Levashef Harbor. Krenitsyn and Levashev remained here during the winter of 1768–1769. A large part of this bay is found in the atlas accompanying the voyage of the *Rurik* (Kotzebue). And a description of it is in Lütke's *partie nautique*, p. 280, with the profiles VIII, IX, X, and XVI.

63. It is possible that this earthquake had something to do with those conspicuous fluctuations of the barometer (and of the magnetic needle?) which Roquefeuil observed from the eighteenth to the twenty-first of March, 1818, at $150°$ Long., and $18°$–$25°$ N. Lat. (cf. Roquefeuil, *Voyage autour du monde*, vol. II, p. 3).

64. This, as well as Langsdorff's report above, is an error. It was probably meant to mean, "in smoke." Characteristic of the Makushin Volcano is that it does not erupt in flames, this according to all other reports, including Cook's (from 1778).

65. According to L. von Buch, this is genuine trachyte [fine-grained, extrusive rocks having alkali feldspars] with columns of hornblende and yellow feldspar [orthoclase?].

66. In the mountains behind Mokrovskoi [Pumicestone] Bay there is, according to Veniaminov, I, p. 171, a lake. In the middle of it there is a small island, on the eastern, rocky bank of which amber is found.

67. Postels, p. 18. Detailed indications of the location of the find are missing in his index of the types of mountain ranges. Langsdorff, II, p. 29, merely states: "The mountains, some higher, some lower, are transected with irregular valleys where common clay is found, and soil, which is washed from the mountains."

68. According to Veniaminov. But Lütke mentions Tulikskoi Volcano.

69. According to Veniaminov. Lütke calls it Stepanof Bay.

70. It would not be without interest, according to the Archive of the North American Company, to ascertain if Kriukov, whom Kotzebue met on Unalaska in 1817 (cf. also Choris, *Voyage pittoresque*), had been on Umnak twenty-one years earlier.

71. According to Baranov, a black substance was also brought up to the surface. It resembled soot. And there was a large amount of small, burnt stones (slake, lapilli).

72. [Cf. Merck, *Siberia*..., pp. 64, 65].

73. Because of the difference of the reports (cf. also the newspaper report in the *New Geographic Ephemerals* III, p. 348), Hoff (II, p. 415) was inclined to assume the formation of two new islands. But that is an error.

74. These data deviate in essence from those of Admiral Krusenstern, which have been used in Berghaus's *Geography and Ethnography*, II, p. 759. According to Vasil'ev, chief of the two ships, *Otkrytie* and *Blagonamerennyi*, the circumference, as Krusenstern states, is supposed to be four miles, and the elevation 2,258 ft. (350 *Tois.*). But the two ships parted in May of 1820; and Vasil'ev went to Kamchatka. I have not been able to ascertain whether he could have made these observations perhaps in 1819 (cf. Appendix II, 1818–1822).

75. The location of the same is not indicated accurately enough. The same is true of the solitary cliff, mentioned by Stein, which was known to the Aleuts long before the genesis of St. J. Bogoslof. And from there a reef extends toward Umnak. It probably lies west of St. J. Bogoslof.

76. That was the year it emerged above the water's surface; but volcanic phenomena were observed at its present location as early as 1795 (Langsdorff in Krusenstern, III, p. 142).

77. [Midway?—Tr.]

78. We would gladly leave such consideration to seamen. Neither do we believe that such and similar opinions (cf. Veniaminov, I, pp. 301–315) and proofs for the existence of yet unknown islands in the district of Unalaska are satisfactory. But we will be held blameless, when taking into consideration that we do not aim at the discovery of new islands, but only at the geologic interest of the object. It is, incidentally, Lieutenant-Commander Kashevarov's opinion, that much about the above-mentioned tale is untrue. He intends to publish a statement to the effect. Erman (III, p. 28) tells of a story by a later discoverer of Unalaska, which is very reminiscent of the report before us.

79. According to Sarychev's entry from 28 May 1792 (II, p. 178 with profile; p. 80, two profiles), there is a mountain on the NE end of the island which once belched fire. But it has not been burning for a long time. We call it Sarichef Volcano.

80. These were used for an analysis. They yielded only Na, and not a trace of K or Ca content.

81. According to Lütke (*p.n.*, p. 313), a second Sand Bay (Peschannaia Bukhta) is located on the southern side of Korovin Bay, a little toward the interior of the same. On the eastern side of this Sand Bay, which opens northward, enormous masses of trees are layered between the rock. Their wood has a grey color. It burns very slowly and is not as good as hard coal. But it is believed that hard coal will soon be found here, because of the occurrence of the wood in this place.

82. The predominant albite strata of this type of rock were also analyzed.

83. Pallas's work reveals in Busching's *Mag.* XVI, p. 257, that Tolstykh first provided these data. They are word for word the same (cf. furthermore, Pallas, *N. B.* I, p. 297, and Schloezer's *Neue Nachrichten von den neuentdekten Insuln zwischen Asien und Amerika*, pp. 64–67).

84. [Merck, *Siberia…*, p. 161, mentions a mountain, but no steam.]

85. [According to Merck, the *Slava Rossii* did not leave Kamchatka until after May 1, 1790. Tr.]

86. Lütke, *p. n.*, p. 335, did not know that Iakovlev's voyage had already been described in German by Pallas in 1781. The same essay, however, was known to him from *Sibirskii Vestnik Spaskago*, 1822, XVIII, kn. 4, from which the map, here attached, was also taken.

87. [Klafter—German measure of length: fathom (which applies here), or volume: cord (a cord of wood).]

88. [Middle German measure of distance or length, used by miners (cf. Duden).]

89. In 1780 Pallas remarks concerning this statement: "According to most recent reports, the amount of copper is not nearly as large. Most of it is transported away by our seamen, the prohibition notwithstanding. Thus only small pieces, the size of beans, can now still be collected." According to Mr. Voznesenskii's reports, copper can presently still be found in larger quantities.

90. Pallas is correct when he regrets in a footnote that Coxe did not adhere to his original narrative, but added a few remarks instead. The original came into his hands by way of the Empress Catherine the Great. Now it can probably no longer be found.

91. "These rocks will readily be recognized as an integral part of the oldest sedimentary formations of Kamchatka, if it is taken into consideration how short the distance is to Bering Island, which consists of cliffs of graywacke, as well as the distance of both islands to the nearest point on the east coast of the peninsula on one hand, and the westernmost Aleutian Islands on the other." It is surprising how easily Professor Erman draws his conclusions. The coast of Kamchatka opposite Bering Island, and that island itself (see below), are as good as unknown so far as their geologic conditions are concerned. The same can be said concerning the copper ores on both sides of the peninsula. Erman says (p. 539) that the ores are probably deposited on tallowy [greasy] shales. If that method is applied, then the sulfur found on Copper Island, by analogy, could be sediment from hot springs, such as are found on the Ozernaia River (outflow of the Kurilian Lake, l.c., I, 3, p. 526). Thus, hot springs could also occur on Copper Island.

92. Neither do I see any indication anywhere that the old Russian verst, certainly smaller than the new verst, could have been shorter by one half.

93. According to Lütke, 2,200 ft. in the southern half of the island.

94. It seems to be a natural cause, that weathering and receding of the surrounding type of mountain exposes such a core progressively more (L. von Buch's Theory Concerning the Rising of Islands, probably provides the best explanation. Author).

95. Steller did not visit the higher mountain ranges on Siberia's borders. In the promontories and mountains of medium elevation, where valleys are washed out by floods, brooks, and springs, Bourguet's rule certainly applies.

96. Probably Sivutchy and Toporkof Islands. There are also small, barren rock islands, such as the Ari Rocks, near the settlement, and then the so called *otpriadyshi* (masses of rock that have collapsed from the shorewall into the sea) mostly on the west end of the island. (Author)

97. It is noteworthy that during the first visit to St. Paul's Island, the scabbard of a sabre, a lime-pipe, and a fireplace were found in a bay on the SW side (Veniaminov, I, p. 132). Perhaps that was Zaikov's landing place.

98. According to a piece from the analogously composed Steffen's Island in Norton Sound, it is probable that the pockets of olivine within the basaltic lava enclose hornblende as well. When the lava is very porous, one finds feldspar, hornblade, and mica burned in a blue clay (cf. "The Phenomena at the Tigil River" [west coast of Kamchatka], in Erman, I, 3, p. 154; and the Island of St. Paul).

99. We take this opportunity to list here an overview of the most important works which treat the geologic conditions of these regions generally or specifically:

Hoff (K.E.A.von) *History of the Alterations in the Earth's Surface*, 2 volumes, 8°, Gotha,1824. Vol. II, The History of Volcanos and Earthquakes, pp. 411–415.

Buch (L. von) *Physical description of the Canary Islands*, 1 volume, 4° with atlas in f., Berlin, 1825, pp. 379–399.

Hofmann (E.) *Geognostische Beobachtungen gesammelt auf einer Reise um die Welt* [Geognostic observations, collected on a global circumnavigation], 8°, Berlin, 1829.

Lütke (F.) *Voyage autour du monde...1826–1829*, 3 vol., 8°, Paris,1835. Vol. 3, Geognostic notes by A. Postels. *Partie nautique*, 4°, St. Petersburg,1836. Cf. also Berghaus (Dr. H.) *General Geography and Ethnography*, 2 vol., 8°, Stuttgart, 1837. Vol. II, pp. 735–745.

Veniaminov (I.) *Zapiski...*, 3 vol., 8°, St. P., 1840. (Veniaminov's description of the District of Unalaska, 3 vol., 8°, St. Petersburg, 1840 [English translation published in 1984 as *Notes on the Islands of the Unalashka District*]. The first volume is important here. It contains a topographic description of the Fox Islands, the western half of Alaska and the Pribilof Islands, furthermore a wealth of natural historic observations. The second volume contains predominantly ethnographic, linguistic, and zoologic information. In the third volume are observations about the Aleuts of Atka and the Kolosh [Tlingit].) Cf. Erman's *Archive*, vol. II, 1842, pp. 459–495 (Veniaminov's work in excerpt).

Stein (F.)...*Vulkanicheskii ostrov Sv. Ioanna Bogoslova*, 4°, St. P., 1825 (Stein, Th., Commentaries to the Depictions of the Volcanic Island, St. John the Theologian or Agashagokh, St. Petersburg, 1825, four pages). Cf. also Krusenstern, *Mem. Hydrogr.*, St. Petersburg, 1827, p. 97.

Stein (F.)...Aleutskikh ostrovov v *Trudakh IMI Mineralog. Obshchestva*, St. P., 1830, 8°, pp. 375–390. (Stein, On the Period of Development of the Aleutian Islands in Transactions of the Mineralogical Society, St. Petersburg, 1830).

Buckland, Wm., in Beechey's (Capt. F.W.) *Voyage to the Pacific and Bering Strait*, 2 vol., 4°, London, 1831. Vol. II, pp. 593–613, Appendix: On the Occurrence of the Remains of Elephants and Other Quadrupeds in the Cliffs of Frozen Mud in

Eschscholtz Bay within Bering Strait and in Other Distant Parts of the Shores of the Arctic Seas.

Buckland, Belcher, and Collie in: *The Zoology of Capt. Beechey's Voyage*, London, 1839, 4°. Appendix, pp. 156–180. Geology.

Girard (H.) in Erman's *Archive*, vol. III, p. 544. Descriptions of Several Petrifications in the Aleutian Islands.

Blaschke (E.) *Topographia medica portus Novi-Archangelscensis*, Petropoli, 1842, 8°, p. 82.

Wrangell (F. von) *Statistische und etnographische Nachrichten über die russischen Besitzungen an der Nordwestküste von Amerika* [Information about the Russian possessions at the northwest coast of America], in Baer's and Helmersen's *Beiträge zur Kenntniss des Russischen Reiches* [Contributions to the knowledge of the Russian realm], vol. I, 1839, 8°. P. 1, Short Statistical Survey of the Russian Settlements in America [English tr. by Mary Sadouski, Kingston, Ontario, 1980]; p. 137, Excerpt from the diary of the Midshipman Andreas Glazunov; p. 161, Notes on the Copper River after Klimovskii's journey, 1819, and Khlebnikov's notes; p. 168, notes concerning two high mountains on the west coast of Cook's Inlet, and concerning the effect of the subterranean fire on the island of Unimak; p. 315, on the earthquake on St. Paul Island, from Chittshinev's [Chichinov's] diary; pp. 323 and 325, Notes on Alaska and St. Paul.

A survey of the geological results of the voyages of Franklin, Richardson, Beechey, etc., we find in *Edinburgh Cabinet-Library*, 1831, vol. I, pp. 443–468, and the rest of the geologic quotations from the various travelogues in the descriptive part of this work, as well as in Appendix II.

100. Veniaminov mentions in his description of the region of Unalaska, St. Petersburg, 1840, vol. I, p. 5: "Presently there are nine volcanos on the Fox Islands and Alaska, one on Yunaska, one in Akutan, one on Akun, three on Unimak, and one on Alaska. With exception of the Shishaldin on Unimak, they only smoke and have never spouted flames." Cf. below, pp. 116–121.

101.

FROM	TO	DISTANCE (NAUT. / GEO. MILES)
Wrangell Volcano	Iliamna Volcano	380/95
Iliamna Volcano	Pavlof Sopka	390/97.5
Pavlof Sopka	L. Sitkin	735/183.75
Little Sitkin	Cape Kamchatskoi	580/145
	Total distance	2,085/521.25

102. We want to mention in this regard the volcano recently mentioned by Veniaminov in 56° N. Lat. and 157°–159° W. Long., and the volcanic phenomenon observed by Don Salvador Fidalgo in 1790 at the northern interior side of Chugach Bay in 60°54' N. Lat. (cf. Appendix II, 1790).

103. Concerning Mt. Hooker and Mt. Brown (52°–53° N. Lat. and 118°–119° W. Long.), we do not know if they are of volcanic nature. On the whole, we lack more accurate information on this part of the Rocky Mountains. Only in the *Travels* of Lewis and Clark, p. 214, do I find an entry which indicates possible earthquakes: "Since our arrival at the falls (of the Missouri, 47° N. Lat. and 50°30' Long. west of Washington, June and July

1805) we have repeatedly heard a strange noise coming from the mountains in a direction a little to the north of west. It is heard at different periods of the day and night, sometimes when the air is perfectly still and without a cloud, and consists of one stroke only, or of five or six discharges in quick succession. It is loud and resembles precisely the sound of a six pound piece of ordnance at the distance of three miles. The Minnetarees frequently mentioned this noise like thunder, which they said the mountains made; but we had paid no attention to it, believing it to have been some superstition or perhaps a falsehood. The watermen also of the party say that the Pawnees and Ricaras give the same account of a noise heard in the Black mountains to the westward of them. The solution of the mystery given by the philosophy of the watermen is that it is occasioned by the bursting of the rich mines of silver confined within the bosom of the mountain."

104. The Kambalinaia Sopka of Chirikov, the Kasheleva or Opalinskaia of Krusenstern and Erman's First Volcano (51°30′ N. Lat., 154° 56′ Long. east of Paris) are one and the same mountain. Its location, according to Chirikov and Krusenstern, is 51° 22′40″ N. Lat. and 156° 58′20″ Long. east of Gr. Cf. The Map of Kamchatka, shortly to be published by the Hydrographic Department of the Naval Ministry.

105. Cf. Berghaus, *Physical Atlas*, Section III, Map 9, with text pp. 54 and 56.

106. [Note from p. 423:] Wherever Wrangell's Glaciers along the Copper River are mentioned, we mention in addition that Captain Belcher's presumed glacial formations at Glacier Bay (cf. pp. 16–18) ought to be more closely investigated, as well as those already observed by Malaspina (59°59′ N. Lat.) in Yakutat Bay to the harbor Desenganno [Disenchantment Bay] (Vancouver, V, p. 67; Humboldt, *Nouv. Esp.*, I, p. 349). This because under 61° N. Lat. Mackenzie, e.g., found the Great Slave Lake still partially frozen in June.

107. Cf. Berghaus, *General Geography and Ethnography*, 1837, vol. II, p. 729–754. In that place the above information is found, according to Erman, while in his *Reise um die Erde*, I, vol. 3, p. 558, the greatest occurrence of activities of the Kliuchevskaia Sopka takes place in the years 1729, 1737, 1796, and 1829. We regret that the fourth volume of that voyage, which is to treat the regions we have here described, has not yet been published. We hope that our work has come out early enough to be of use to that famous traveler for a compilation of the volcanic phenomena of Kamchatka, the Aleutian Islands, and Alaska. Thus, we hold ourselves in abeyance in this regard, knowing that Professor Erman has certainly more material available to him. In this place only one comment: Krasheninnikov assumes that the recurrence of the eruptions of Kliuchevskaia Sopka happens every eight to ten years. Erman, too, seems inclined to believe that. But, according to available older reports and recently arrived observations, we seem to find intervals of four, five and six years indicated. The years are as follows: 1727–1731; 1737; 1762; 1767; 1773 (Avacha Sopka); 1796; 1829; 1842; 1846. But since all these reports are far from giving us a comprehensive picture, it is perhaps best to refrain from assuming such a regular periodicity.

108. Erman submits evidence of a relationship of alternating activity of the Kliuchevskaia and Avacha volcanos (*Reise um die Erde*, I, 3, p. 537).

109. At the end of the last [eighteenth] century the Kambalinaia and Kliuchevskaia Sopkas were especially active; only neither of the dates are exactly known. 1795, in fall, on

Unimak, at the Pogromnoi, horrendous eruption and fallout of ashes. 1796, on May 7, St. John Bogoslof; Kliuchevskaia Sopkai Edgecumbe? 1796, in November, the volcano of Pasto begins to smoke. 1797, the fourth of February, normal standard, destruction of Riobamba. 1797, the twenty-seventh of September, eruption on the West Indian Islands; Volcano of Guadalupe. 1797, the fourteenth of December, destruction of Kumana. 1798, the ninth of June, lava eruption at Cahorra Volcano on Teneriffe (Canary Islands).

110. Some call them, perhaps somewhat daringly, the North American Urals.

111. Almost exactly under latitude 52° there rise the volcanic islands, from Little Sitkin to Great Sitkin. Only Goreloi and Sitignak are under 51°45′ N.Lat.

112. Except for Amlae [Amlia], which extends from W to E, and is similar to Kiska, which lies an equal distance from Tanaga and extends, in contrast to the other islands, SW to NE.

113. Cf. the compilation of the summit elevations of these regions on Map II.

114. The coast of the Baltic Sea, the west coast of Italy, the region of Cush between Indus and Gudgerat, and the coast of Arracan (Island of Reguain).

115. L. von Buch (Treatise of the phys. class of the Academy in Berlin, 1812–13, p. 143, and 1818–19, p. 60; or Hoff's *History of the Alterations of the Earth's Surface*, II, p. 415) says that it is not a mere mass of expulsed slake and fragments like, for instance, Monte Nuovo. Instead, it seems to be one of those peculiar bubble-like elevations, examples of which so often appear in so many volcanic regions [cone vs. shield volcanoes].

116. Map II is therefore only to be viewed as a first attempt of a geologic sketch. By no means can there be a claim of exact limits of the existing types of rock.

117. Cf. von Helmersen, Göppert, and Count Keyserling, in A.Th. von Middendorff's *Sibirische Reise*, vol. I, Th. I, pp. 196–274.

118. In Prof. Erman's travelogue we miss most of all indications of observed conditions of stratification. The vivid imagination of our ingenious traveler substitutes this with inferences concerning the compositon of points far removed from one another, concerning the age of their genesis, etc., on the basis of rocks, defined in part by our certainly most excellent petrographer, G. Rose, or according to designations by specific weight and some analyses. The geologic portait of Kamchatka seems therefore to be for the most part hypothetical. About twenty years have passed between Erman's journey and the publication of the third volume of his report. From the geological point of view, we would, instead of the latter, rather have welcomed a coherent, fairly complete work. Should our attempt at a dense survey of the data dispersed in the mentioned work not comply with Prof. Erman's wish, as is to be expected, then he must ascribe the blame to himself. It served no purpose to wait any longer, although we would rather have been spared the troubling work of compilation.

119. This definition of iron and marl deposits at Cape Omgon ("seen through the telescope," Erman, *Reise*, I, 3, p. 125), at the Tigil (l.c., p. 149), and Kultuk (l.c., p. 151: "impressions of several kinds of tree-leaves") is unreliable, because it is based on only two dicotyledons (*Juglans* and *Carpinus*), one *Modiola jugata*, and *Anadonta tenuis* (Girard). But Erman goes on to say that, considering these fossilized remains, the west coast of the peninsula should be taken for the eastern shore of a freshwater lake, which must have reached toward the other side (!) as far as the spit of land between Penjinsk and Iginsk

Bays. The *Modiola jugata*, finally, causes the same author to assume (l.c., p. 152) the occurrence in the past of a transformation from fresh water to an inundation by the sea.

120. Erman's report, I, 3, p. 204. "Through the deposits of the chalk formations, i.e., after fresh water had covered them (cf. the last footnote), augitic and trachyte porphyries emerged, which at the Tigil and Pallan Rivers formed porphyry-like amygdaloids. They are of simultaneous origin with the marekanean trachyte and marekanite rock, which must be considered metamorphous types of rock (l.c., pp. 94 and 197). This outcropping, however, must have happened prior to the inundation with seawater of the land east of there, which then deposited layers of limestone containing *Tellina dilatata*, *Natica aspera*, *Crassatella*, *Venus*, *Nucula*, *Buccinum*. Consequently, the origin of these formations must belong either to the epoch of the younger chalk period or that of the older Tertiary formations." But according to what we know about these formations, they seem to be analogous with deposits on the Aleutian Islands. Therefore, they seem to belong among the youngest Tertiary formations.

121. Also deposits of brown coal with contents of amber at the Sedanka River, a tributary of the Tigil River, with remnants of conifers, imprints of grasses, and petrified pieces of branches and stems of a deciduous kind of tree (l.c., p. 211). At the bottom of page 213 it is furthermore stated: "That black soft coal, very similar to the Sedankean coal, which also contains amber, is found in the proximity of the Arctic Sea and especially at the estuary of the Yenisei River, 73° N. Lat., 80° Long. east of Paris or 77° west of Sedanka, on one hand, and on the other hand along the slopes of the Aleutian volcanic chain near its proximity to the American continent, on Alaska and Unga, 56° and 55°05' N. Lat., 200° Long. east of Paris or 45° Long. from Sedanka, as well as on Sitka. This only proves for the time being, that a similar vegetation and a similar demise of interconnection of the basins, however, where those masses, so far removed from each other, were deposited, can neither be so patently presumed, nor should it be presumed to be a priori possible." Why these similitudes: they only prove that 77° W and 45° E of Sedanka the geognostic composition of Asia, of the Aleutian Islands, and of Alaska is little known, and that the available material is not adequate to compare the mentioned deposits with one another (l.c., p. 113). The statement continues: "At any rate, the timbers in the deluvial deposits along the Lena River below Yakutsk, and the wood-mountains of New Siberia, or Kotelnoi Ostrov and other islands in the Arctic Sea, which were deposited together with the remnants of pachyderms, were distributed by a far later flood than the Tertiary coals of Kamchatka and the Aleutian Islands" (cf. I, 2, p. 260). According to the geognostic sketch of Northern Asia, Jura [Jurassic] formations are widely distributed on Kotelnoi Ostrov, at the estuary of the Lena River, etc. Mr. Erman could have mentioned the difference of these informations. Wood-mountains and bones of mastodons occur also in the Aleutian Islands (Unalaska), on the Pribilof group, and at the east coast of the Bering Sea. How a flood could have distributed the Tertiary coals on the Aleutian Islands, Alaska, and Chugatsk, does not make sense to us.

122. Trachyte and lava from Baidara Mountain (l.c., p. 228) to the Povorotnaia Sopka (p. 283). "A grey, trachyte-like mass seems to have arisen in innumerable locations of that zone occupied by the Kamchatkan Middle Range in ribs that diverge and fall off in those

places where they run together. But thereafter, in epochs hard to differentiate from the time of their emerging, they seem to have been broken apart by lava and loose pumice stones, which no longer seem to differ from them, much like the molten rock from the same in its original condition."

123. The star-shaped grouping of the ridges or table-shaped ribs surrounding the Shivelich Volcano is typical for that mountain (p. 295). The origin of all of its parts obviously occurred all at once (p. 297). The star-shaped groupings of the andesite table welled up as a soft mass of mountain material, which opened its source as it congealed and filled the trenches. Similarly shaped conditions as those existing at the Shivelich Volcano, can be found in the buttress-shaped pillars at the prismatic summit of Mt. Chimborazo, and also in the folds and ribs of Mt. Aetna. But the augite-porphyry and dolerite of the Chimbarazo Volcano are not penetrated by any crater. And the volcanos Tunguragua, Pinchincha, and Antisana terminate in andesite. But the lava-volcanos of Kamchatka (Kliuchevskaia Sopka) stand in augite-porphyry. And wherever no crater formation is present, or where it cannot be proven to exist (Shivelich, p. 298), there the rock is andesite. The Shivelich Volcano, with its total lack of lava or open craters, and with andesite as its major rock content, offers, therefore, an analogy rather with the Caucasian cone-mountains (Ararat), which are interpenetrated with augite and masses containing labrador (Kasbeck, p. 299). Craters have become known in Trans Caucasia, which are filled with lava up to the rim. And specimens sent by Voznesenskii speak against a lack of lava at Mt. Shivelich. Erman's own data suggest that Mt. Shivelich was once an active volcano. On p. 479, for instance, he states: "The Iupavona Sopka (55°55′ Long. and 156°8′ Lat. East of Paris) because of its expulsion of smoke and its subterranean rumbling beneath its ridges, has always been compared with Mt. Shivelich and is therefore counted as definitely among the fire-mountains." Thus Erman, l.c., p. 285 (or the footnote on the preceding page), expresses clearly that lava and loose pumice broke immediately following or almost simultaneously with the steam, from fissures very close to the summit of Mt. Shivelich, in very small amounts and without the formation of any lava. "The emergence of this slake has been, at any rate, a rather secondary phenomenon. It stood in a very remote connection with the forces which once formed that mountain." Concerning the Tolbachinskaia Sopka, the assumption is stated, p. 405, that it has a core of andesite, and that its base is most closely enveloped in augite-porphyry, just like the mountains at the southern point of the peninsula.

124. According to p. 412 of the report, the genesis of the oldest mountains, i.e., the Chapinskaia Sopka, the Kronotskian cone, Mt. Shishel, and other volcanic cones with flat summits and blunted ridges, as well as the other andesite mountains of the peninsula, falls into the time after the opening of the present valleys from the diorite surface and metamorphic shales. But the sketch reveals that south of Nizhne-Kamchatsk toward the Kronotskaia Sopka and farther toward the southern point of the island [peninsula?], the geologic compositon is that of the middle range.

125. Our knowledge of the structure of these volcanic cones is too inadequate to allow us to employ Mr. Erman's method (III, p. 275) for calculating the transformations which result from the weathering of the rock plateaus and ridges into rounded forms.

126. We believe we ought to quote here a word from Mr. Naumann (*Textbook of Geognosy*, I, p. 641), our thorough and acknowledged critic: "The very uncertain concept of the andesite can only be approximately ascertained, for the time being, by its peculiar outer habitus, and by the geognostic characteristics of the deposits. No volcanic type of rock exists in a larger variety than andesite. It traverses all the gradations from silicon-rich trachyte-porphyry almost to dolerite." The new designations of rock types, which happen not rarely of late, prove the beginning of more precise defining, as well as the inadequacy of our present methods of defining. We do not want to indicate that the name andesite, especially as used for certain deposits, is useless. Only the expressions albite-trachyte or Andean trachyte would have been more indicative. We are reluctant to number the trachydolerites found on the islands between Asia and America among the andesites. We do not believe it is necessary to honor these characteristic types of rock with new names like Aleutite, Beringite, etc. And the right to such christenings ought to be reserved for those geologic conditions of the Andes on the whole too little known to make such easy assumptions and opinions concerning the distributions of certain types of rocks, of which they are composed. The attempt to gain new insights everywhere is certainly worthy of recognition, so why not also in the classification and naming of rock types? Andesite, however, can with very few exceptions, dissolve in trachydolerite. Therefore, we have not adopted the name andesite on our Map II.

127. Nevertheless, the indications concerning the granite-like rock deposits derive from people who are not geologists. Thus, they might at times be mistaken.

128. Ingenstrom's claim that Goreloi Island, Tanaga, and Kanaga reach an elevation of over 9,000 ft. must be doubted.

129. The rest of the locations of the finds, and especially those indicating the deposits and quality of the different types of coal, are too little known to allow us to assume knowledge of the same with certainty.

130. In Berghaus's *Annal.*, Vol. VI, Erman states that Kamchatka's major elevation is to be placed between the periods of the tallow-like transition ranges and the formation of the brown coal deposits. The elevation of the mountain passes is supposed to diminish from the east coast toward the west coast, and likewise toward the Komandorskie Islands, and on those islands as well. According to Eli de Beaumont, Kamchatka was lifted up after the settlement of the Jurassic formations and before the displacement of the lower chalk (Neocomian and Greensand).

131. Cf. Cochrane versus Burney, 1821, in *The New Comprehensive Geographic and Statistical Ephemerides*, vol. XVII, Weimar, 1825, pp. 385–467; and Kotzebue 1815–1818, III, p. 157.

132. In the manner, for instance, of calculating 9,000 or 4,500 years for ninety Aleutian Islands.

133. The Komandorskie Islands were uninhabited when they were discovered, and without traces of earlier habitations.

134. Cf. also Berghaus, *Physical Atlas*, and the commentaries on the same, pp. 195–198.

135. I take this opportunity to point out especially, how crucially advantageous has been for me the excellent work by von Middendorff, *Beiträge zu einer Malacozoologia Rossica*, St. Petersburg, 1849, 4°, in my effort to describe these Tertiary fossils. The same work

also treats of the mollusks of these oceans. The author just mentioned was guided by an especially excellent method of handling the measurement relationships, which resulted in clarification of older, unclear and confusing definitions. These dimensions cannot be applied to the seldom completely preserved fossils, to the same extent that they are applied to living species. Nor can they be a direct major factor toward arriving at classifications. But it does make sense that the only way we can find our way out of the maze of petrifactological [paleontological] studies, particularly those on younger formations, is by a strict adherence to the living fauna. Thus, indirectly that method is of immense value. Without Middendorff's work the description of our bivalves would merely have contributed to the confusion in the classification of Tertiary mollusks.

136. *A Memoir of Sebastian Cabot, with a Review of the History of Maritime Discovery.* London, 1831, 8°.

137. On the Spanish voyages before Malaspina, 1790, the following publication is most informative: Relacion del ultimo viage al Estrecho de Magallanes, en los annos de 1785 y 1786.—Extracto de todos los anteriores, des de su descubrimiento, impressos y Mss. Trabajada de Orden del Rey. Madrid 1788. Por la viuda de Ibarra, etc., 4°, with maps.

138. Cf. J. L. Forster, and Scoresby, Chron. enum. According to Fleurieu, in Marchand's *Voy.*, I, p. iii, Cortez himself participated in the expedition on the *Capitane* and discovered California. Thus he contradicts Forster, but without properly motivating his statement. And he seems to have no more sources available to him than does Forster, viz.: Antonio Herrera, *Description de las Indias*, Ambères, 1728, fol., Decad. V, Lib. 8, Cap. 9, 10; Decad. VII, Lib. 6, Cap. 14; Venegas, *Hist. de la California*, p. 124; Lorenzana, *Hist.*, p. 522; Robertson's *History of America*, Lib. V.

139. In the publications of the Archeographic Society at St. Petersburg (*Dopolnenie kartam istoricheskim izdan. Arkh. Kommissieuiu* [Supplement to the historical maps published by the Archeographic Commission]) there will soon appear a manuscript, found not long ago, which deals with the voyage of Dezhnev and Motora. Cf. the course of these voyages on Müller's map from 1758.

140. Müller (in his *Sammlung Russ. Geschichte*, pp. 81–85) writes "Kobelef," Pallas (*N. B.* IV, p. 112) writes more correctly "Kolessof." His source is Adelung, *G. d. Schiff.*, pp. 525–541, where this part is mostly based on Krasheninnikov's description of Kamchatka, pp. 291–301, and treated more accurately than in Müller's work.

141. Pallas (l.c., Appendix) writes erroneously "Lukin," because in *Polnoe sobranie zakonov* [Full code of laws] is written: "Luzhin" Cf. also Müller's *R. G.*, p. 109, and Adelung's *G. d. Schiff.*, p. 546.

142. Concerning this and the following voyages of promyshlenniks, cf. in addition to Müller, Coxe and Busching's *Mag.* XVI, pp. 235–281; Berkh, *Khronol. istoriia*; and Veniaminov, *Zapiski ob ostrovakh Unalashkinskago otdela*, St. P., 1840, I, pp. 113–133.

143. [The naturalist on board was Carl Heinrich Merck. Cf. his *Siberia and Northwestern America, 1788–1792. The Journal of Carl Heinrich Merck, Naturalist with the Russian Scientific Expedition Led by Captain Joseph Billings and Gavriil Sarychev*, translated by Fritz Jaensch, edited by Richard A. Pierce (Kingston, Ontario: Limestone Press, 1980).]

144. [His account is translated in *Russian Exploration in Southwest Alaska: The Travel Journals of Petr Korsakovskiy (1818) and Ivan Ya. Vasilev (1829)*. Translated by David H. Kraus, edited with an introduction by James W. VanStone. Fairbanks: University of Alaska Press, 1988.]

145. [Since translated in *A. F. Kashevarov's Coastal Explorations in Northwest Alaska, 1838*. Edited by James W. VanStone. Fieldiana: Anthropology, 69. Chicago: Field Museum of Natuaral History, 1977.]

146. [For a translation of his account see *Lieutenant Zagoskin's Travels in Russian America, 1842-1844*. Translated by Penelope Rainey, edited by Henry N. Michael. Toronto: University of Toronto Press for the Arctic Institute of North America, 1967.]

Index

➤➤ ◄◄